RAINFOREST

Sara Oldfield

Photography by Bruce Coleman Collection

Foreword by Mark Rose,
Executive Director of
Fauna & Flora International

The MIT Press
Cambridge, Massachusetts

Published in the USA by The MIT Press
First published in the United Kingdom by
New Holland Publishers (UK) Ltd

Library of Congress Cataloging-in-Publication Data
Oldfield, Sara.
  Rainforest / Sara Oldfield ; photographs by Bruce Coleman
Collection ; foreword by Mark Rose.
    p. cm.
  ISBN 0-262-15106-5 (hc.: alk. paper)
  1. Rain forests.   I. Title.

QH86.O43 2003
578.734—dc21

2002029559

Publishing Manager: Jo Hemmings
Project Editor: Lorna Sharrock
Editorial Assistant: Daniela Filippin
Designer & Cover Design: Becky Willis at Design Revolution
Copy editors: Tim Sharrock, Julia Cady
Production: Joan Woodroffe

Reproduction by Pica Digital Pte Ltd, Singapore
Printed and bound by Tien Wah Press (Pte) Ltd, Singapore

Front cover: Toco Toucan
Back cover, Page 1: Red-eyed Tree Frog
Page 2: Rainforest flower
Page 4: Caterpillars on leaf in Costa Rica
Page 6: Red-eyed Tree Frog, Mountain Gorilla with baby,
Chamaeleo oustaleti in Madagascar, Chequered Swallowtail
Butterfly, Rainforest in Sumatra
Page 7: Resplendent Quetzal, St Lucia Parrot, Flower of
Mamorana Tree, Jaguar, Rainforest in New Zealand
Page 8: Jaguar

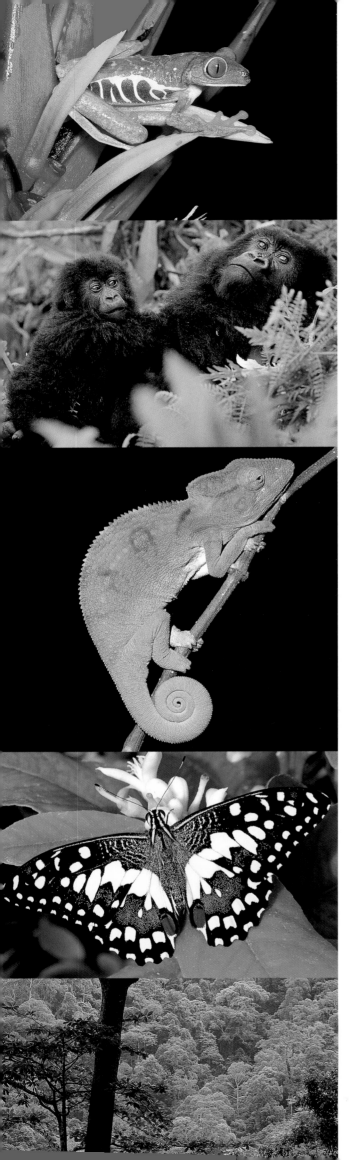

Contents

Foreword 8

Chapter One

Introduction –
Rainforests around the World 10
Rainforest Diversity 12
The Human Factor 13
A Vanishing Habitat 18
The Future of Rainforests 22
Managing the Rainforests 26

Chapter Two

Africa –
Gorillas in the Mist 28
Rainforests of West Africa 30
Rainforests of Central Africa 34
Rainforests of East Africa 40
The Future for African Forests 42

Chapter Three

Madagascar –
A World Apart 44
Natural Riches 46
Man in Madagascar 48
Protecting Madagascar's Heritage 50
Lemurs in Danger 52
Plants in Peril 54

Chapter Four

India and South East Asia –
Forest Treasures of the East 56
Asian Rainforests 58
India 60
Vietnam 62
Cambodia 66
Laos 68
Thailand 70
Malaysia 72

Chapter Five

Indonesia and the Philippines –
Islands of Glory 80
Indonesia 82
Wildlife Under Threat 90
The Philippines 94

Chapter Six

Central America – A Rainforest Corridor 100
Mexico to Panama 102
A Wealth of Plants 104
Endangered Animals 106
Eco-tourism 108

Chapter Seven

The Caribbean – Fragile Fragments 110
Island Forests 112
Montserrat and Dominica 113
Jamaica 114
Cuba 116

Chapter Eight

The Amazon – The Place and its People 118
Forest Ecology 120
Man and the Rainforests 122
Endangered Species 126
Conservation Measures 128
People of the Forests 130

Chapter Nine

Brazil – Atlantic Forests 132
A Narrow Belt 134
Too Late for the Trees? 136
Primates on the Brink 138

Chapter Ten

Temperate Rainforests – Cathedrals of the Natural World 140
The Global Picture 142
Tasmania 144
North America 146

Original Rainforest Areas of the World 150
Rainforest Species of Concern 152
Useful addresses 154
Further information 155
Glossary 156
Index 157
Acknowledgements 160

Foreword

As Executive Director of Fauna & Flora International (FFI) I have been privileged to explore rainforests in Africa, Asia, North, Central and South America. Each time I visit a new rainforest area I am awed by the sheer diversity and abundance of animal and plant life. The Mountain Gorillas of Rwanda, the Resplendent Quetzals of Central America, the birdwing butterflies of Papua New Guinea and the towering trees of Kalimantan are diverse highlights of my rainforest travels. The hospitality of Mayan people in Belize will never be forgotten.

It may be hard to capture the sights, sounds and atmosphere but back home, books, photographs and artworks keep me in touch with the rainforest inspiration.

My travels around the world have also made me acutely aware of the fragility of rainforest ecosystems and the seemingly relentless pace of destruction. Some of the forests I visited twenty years ago are now oil palm plantations or abandoned subsistence farms stripped of vegetation. But I remain optimistic that we can reverse the trend of rainforest loss in the twenty-first century. As one of the major international organizations committed to rainforest conservation, FFI is delighted to be associated with the publication of Rainforest.

The intention of this magnificent book is to condense the rainforest majesty and diversity into a single volume. It also serves to highlight a selection of the rainforest conservation actions, which are currently being successfully undertaken. Of course there is much more that can be done to save the rainforests. I hope that reading the authoritative text of this book or enjoying its captivating images will entice you to visit a rainforest, find out more about the species that live there and support rainforest conservation. Everyone can play a part however far from the rainforests our homes may be.

Royalties from Rainforest will help to directly support the Global Trees Campaign a joint initiative of FFI and the United Nations Environment Programme – World Conservation Monitoring Centre (UNEP – WCMC) to save the world's most threatened tree species from extinction. Even now we can only guess at the total number of tree species within the rainforests but there is abundant evidence of the threats faced by many rainforest trees. For every tree species lost, an estimated 300 associated species will also die out – pieces of the rainforest jigsaw that are irreplaceable. Around the world experts working with local communities are striving to save threatened rainforest trees and their habitats. With your help their efforts will intensify. Please do all that you can to help.

Mark Rose
Executive Director,
Fauna & Flora International

Chapter One

Introduction

Rainforests around the World

Christopher Columbus described the Caribbean islands as
'filled with trees of a thousand kinds and tall, and they seem to
touch the sky. And I am told that they never lose their foliage,
for I see them as green and lovely as they are in Spain in May.'
Columbus shared with travellers throughout history his sense
of wonder at the sheer magnificence of rainforests.
The biological riches of this extraordinary habitat inspired
Darwin, Wallace and other great Victorian naturalists. Yet
rainforests are still poorly understood scientifically.
Nature lovers of the 21st century may be awestruck by the
natural bounty of rainforests, and by the myths and legends
that surround them. We may now have a better grasp
than ever before of just how essential their unique
ecosystems are to the health of the planet.
Yet much about these rich and often ancient
forests remains a mystery.

Rainforest Diversity

Rainforests grow only in areas where rainfall is high throughout the year. Tropical rainforests need high temperatures, while temperate rainforests are found in cooler regions of the world, but persistent moisture is a feature of both.

Scientists recognize and classify many types of forest according to their physical appearance and species composition. The main types of tropical rainforest include:
- *Lowland rainforest,* the densest and most luxuriant type of rainforest
- *Swamp forest,* found inland, in places with poor drainage or seasonal flooding
- *Mangrove forest,* found where land meets sea
- *Cloud forest,* mysterious, mossy forests found at high altitude.

In reality, forest types generally merge, without clear-cut boundaries, except where these have been introduced by the bulldozer and the chainsaw.

Rainforests occupy a mere six per cent of the Earth's surface, but estimates put the number of species they contain at up to 30 million. No one really knows for sure just why tropical rainforests should contain such an amazing diversity of species. It is true, of course, that plant life thrives in the heavy rainfall and uniformly high temperatures of the tropics. The abundance is also partly explained by the stability of the rainforest ecosystem over a relatively long period. But the ecological functioning of rainforest systems is far from fully understood. To catalogue even a fraction of the biodiversity of this increasingly endangered habitat is a race against time. This is just one of the huge scientific challenges that rainforests present.

PREVIOUS PAGE Rainforest at sunrise. Water and sunlight are two of the vital ingredients sustaining lush rainforest growth. Rainforests are peaceful and tranquil places, unless shattered by the activities of Man.

LEFT Mangrove trees, with their characteristic stilt roots, are especially adapted to grow in salt water. Mangrove forests provide the breeding grounds for the marine life on which coastal fisheries depend.

RIGHT At night, the loud chorus of courting male Red-eyed Tree Frogs can be heard throughout the Australian rainforest.

The Human Factor

Rainforests are immensely important to the well-being of people around the world. Altogether, some 350 million of the world's poorest people depend almost entirely on forests for their subsistence and survival needs. For about 60 million indigenous people and other forest-dwelling communities, forests provide everything for daily life – shelter, fruits, vegetables, meat from wild animals, materials for clothing, gums, dyes and medicines. Forests also have aesthetic and spiritual importance to the people who depend on them. About another 200–300 million rural poor depend on the land in rainforest areas. These people may not own land, but clear patches of rainforest for short-term cultivation before moving on. Sadly, all the people who for centuries have had the closest relationship with forests are nowadays the least well equipped to compete with international economic interests for dwindling forest land and resources.

Most of us live far from the world's rainforests, but we rely on their products and services to a surprising extent. Rainforests are important regulators of the world's climate. They act as natural chemical factories, producing 20–30 per cent of the world's oxygen. They are also a storehouse for species and genetic diversity. Rainforests provide us with a huge range of natural products and – both directly and indirectly – a constant supply of economic goods.

Rainforests and the timber trade

Timber is the most valuable resource harvested from the wild. The main fuel in many developing countries, it is also used in both traditional home-building and more sophisticated construction, and is the basis for the international pulp and paper industry. One of the most important commodities in international trade, timber forms a very important part of the export earnings of developing tropical countries. Worldwide, timber exports are worth over £67 billion ($98 billion). The bulk of the wood in world trade comes from temperate sources. The USA, Russia and Canada are the major exporters of logs and sawn wood, while for plywood Finland, Canada and Russia are in the lead. The main tropical source countries are Malaysia, Papua New Guinea and Gabon for logs, and Malaysia, Indonesia and Brazil for sawn wood and plywood.

Fuel wood is another immensely important rainforest product. Although fuel is not measured on the same economic scale as timber, more than two billion people rely on fuel wood and charcoal as their primary source of energy. Imagine the economic equivalent in terms of gas or electricity.

After timber, rattans are the second most important source of export earnings from tropical forests. Most of the 600-or-so species of these climbing palms are native to South and South East Asia. Countries with major rattan industries include the Philippines, China, Indonesia, India, Sri Lanka and Thailand, providing full-time employment for at least half a million people. For the international market, rattans are mainly used for cane furniture. Local uses include the production of mats, baskets and fish traps, as well as dyes and medicines.

The rattan industry relies largely on wild plants. About 90 per cent of the world's supply comes from the wild, and the remaining 10 per cent from plantations in Kalimantan, the Indonesian portion of Borneo. Exploitation and habitat destruction have led to the decline both of major commercial rattan species and of those species that are valuable in local use and local markets.

Medicinal plants

Medicinal plants are still largely harvested from the wild, and relatively few species are cultivated as crops. Rainforests yield a particularly rich diversity of medicinal plants. In many cases, biochemicals extracted from plants have been used as blueprints for the synthesis of drugs, so the natural source material is no longer required. Nevertheless, the USA annually imports over $20-million (£15-million) worth of rainforest plants for medicinal purposes. Important drugs include *Tubocuranin* derived from plant-based curare, used as a muscle relaxant during surgery, and *Curianol*, a Guyanese fish poison used in heart operations. The US National Cancer Institute has identified over 1,400 tropical forest plants with the potential to fight cancer. One such plant is the Rosy Periwinkle, native to Madagascar. Used for generations by tribal healers, it is now used to produce drugs effective against Hodgkin's disease and other forms of cancer.

Aquilaria malaccensis is a South East Asian tree that produces a pathologically diseased fragrant wood commonly known as agarwood or *gaharu*. This provides an oil used to produce incense and products for perfumery and traditional medicines. Virtually all agarwood is extracted from wild trees. Not all trees are infected, but harvesters fell or damage trees in the wild, in their search for the precious infected wood. There are a few areas of plantation in India but even these are under pressure from illicit felling. *Aquilaria malaccensis* has

been recorded as rare or threatened in Bangladesh, India, Myanmar (Burma), Peninsular Malaysia, Singapore and Sumatra. The major importers of agarwood are China, where it is used in traditional medicine, and countries of the Middle East, which buy the best-grade materials.

Fruits of the forest

Some of our most important and popular foods and drinks also have their origin in the rainforests. Major commercial crops such as cocoa, pineapples, bananas, mangos, papayas, avocados and citrus fruits have been developed from wild rainforest species. Some fruits and nuts are still harvested directly from rainforests for supermarket shelves around the world. Rainforest people consume many other food plants, which could become popular on world markets in future if the wild resources survive. Tropical forests are the source of an estimated 2,500 edible fruits. Papua New Guinea alone has over 250 edible fruit tree species, of which around 40 are in cultivation. Kalimantan, the Indonesian portion of Borneo, is the origin of a considerable range of tropical fruits including mango, durian and breadfruit. Local people value many of the

wild fruit tree species and have brought them into semi-cultivation. Traditionally, shifting cultivators did not fell useful trees like these when clearing the forest for agriculture, and they planted forest fruit trees both as a source of food and to attract wild game.

Cocoa originated in the rainforests of the Amazon. The sweet pulp surrounding cocoa beans provided a snack food for the people of the Amazon long before cocoa began to be cultivated as a crop. The cocoa tree has been grown in Central America for at least 3,000 years, and its seeds were used as currency in parts of Mexico until only 150 years ago. As a drink, cocoa became popular in Europe in the 16th century and milk chocolate was developed 300 years later.

Cocoa cultivation spread so rapidly around the tropics that all the cultivated trees were developed from a very few wild

LEFT Over 95 per cent of the world's rubber is produced in South East Asia, with Indonesia and Malaysia being the leading producers.

OPPOSITE TOP The diversity of colourful fruits sold at tropical markets, such as this one in Malaysia, is stunning to behold. Many are still harvested directly from rainforest trees.

OPPOSITE BOTTOM Coffee fruits, Trinidad. Coffee is the most important source of foreign income in Colombia, El Salvador, Ethiopia, Rwanda and Uganda. Wild relatives of cultivated coffee are still found in the rainforest, providing important genetic resources for the future.

ancestors. This lack of genetic variation made modern cultivated varieties vulnerable to disease. The search for wild genetic material to increase both yield and disease resistance has led to exploration and collection of the plant in the Amazon. Unfortunately, however, large areas of the rainforests of Colombia, Ecuador and Peru, the centre of genetic diversity of cocoa, have been lost through petroleum exploration and exploitation and by agricultural colonization in the foothills of the Andes.

Wild coffee is found in the tropical forests of Africa and Madagascar. The most important species in cultivation are *arabica* coffee, from the montane forests of Ethiopia, Sudan and Kenya, and *robusta* coffee, which is native to the Congo basin. Traditionally, coffee berries were gathered from the wild as a snack or used to make coffee juice, some of which was fermented to make coffee wine. Coffee juice remains very popular in parts of Central Africa. Arabs were the first to roast coffee beans and to cultivate coffee trees as a crop. One of the developing world's most important export commodities, coffee is now overtaken in value only by petroleum and its derivatives. Cultivated mainly in tropical mountainous areas with rich volcanic soils, the crop is usually produced by smallholders rather than in large plantations.

In total, there are around 40 species of wild coffee, and these are of immense value as a source of genetic material to improve the cultivated crop. For example, the caffeine-free wild coffee varieties that have been found in the threatened forests of Madagascar have great potential economic value.

A Vanishing Habitat

Rainforests are disappearing at a staggering rate. Every year, according to the United States National Academy of Science, at least 50 million acres (20 million hectares) are destroyed: an area the size of England, Wales and Scotland combined. Virtually all the primary rainforests in India, Bangladesh, Sri Lanka and Haiti have been lost already. The Philippines lost more than half of its rainforest in the 25 years between 1960 and 1985, while Thailand lost 45 per cent between 1961 and 1985. The rainforests of the Ivory Coast have been almost completely logged.

Logging continues to be a major threat. Malaysian, Indonesian and South Korean transnational companies are moving into areas of pristine forest and increasing their logging operations in West and Central Africa, Madagascar, Papua New Guinea and the Solomons, Belize and Surinam. Little thought is given to sustainable logging, to local people or to wildlife. Bribery and corruption are common means of gaining access to the rainforests.

From forests to farmland

Clearing a patch of forest to grow subsistence crops is the only option for many of the world's rural poor. This shifting cultivation or 'slash-and-burn' agriculture is another major cause of rainforest loss around the world. Traditionally, forestland was left to recover after short periods of cultivation, when migrant people moved on to new forest sites. Now, with ever-increasing population pressure on decreasing areas of rainforest land, the fallow period is too short for the forest to recover. New settlers moving into rainforest areas, often as part of government resettlement schemes, lack the traditional knowledge needed to make shifting cultivation sustainable.

Commercial agriculture also plays a major part in rainforest destruction. The spread of oil-palm plantations has, for example, been one of the major recent threats to Indonesia's rainforests. Local people rarely benefit when cash-crop production expands.

The increasing fragmentation of rainforest areas makes them more susceptible to burning. Huge swathes of normally fire-resistant forest have been destroyed or damaged by fires in the Amazon, Central America, West Africa, Madagascar, Indonesia and other places around the world. Most fires in tropical forest areas are started deliberately to clear the land for agriculture or plantation development. Sometimes, fire is used to solve or avenge disputes over land ownership.

LEFT Cracker Butterfly. Butterflies of the rainforest have different types of coloration depending upon where they occur. Some species of the shady forest floor are transparent, whereas brightly coloured orange and blue species fly in the forest canopy.

RIGHT The population of the regal-looking Sumatran Tiger has dropped to no more than 650 individuals in the wild because their forest habitat is being wiped out.

Industrial threats

Mining and oil extraction have had a devastating impact on rainforests in various parts of the world. Fortunately, some major transnational companies are beginning to recognize their environmental responsibilities and the need to minimize the terrible ecological damage they can cause. But this is too late for many areas where the pristine forest and its wildlife have already been wiped out, along with the health and livelihoods of local communities. An area of rainforest at Carajás, in the Brazilian state of Pará, is set to become the world's largest rainforest mining area. Here, a tract of forest larger than England and Wales combined is inexorably disappearing under a huge industrial site with iron-ore mines and smelting plants, aluminium plants and hydroelectric dams.

Oil companies hold leases on nearly all the remaining tropical rainforests, for either exploration or drilling. Their activities have caused significant damage to the rainforests of the Amazon, West Africa and Papua New Guinea. Oil, Ecuador's main source of foreign income, was discovered in the east of the country around 30 years ago and has been

extracted from 6,350 square kilometres (2,450 square miles) of forest. Other areas are being explored.

Endangered species

Each year the list of endangered rainforest species grows longer. Rainforests are home to some of the world's most fascinating and well-loved plants and animals. Many of them are now in trouble. More than 900 threatened bird species are associated with tropical rainforest habitats, over 40 per cent of them with lowland forest and 35 per cent with montane rainforest. The main focus of the conservationists' attention is on such birds and on mammals, also rapidly losing their habitat as rainforests are felled and burned. But tree species, the very fabric of rainforests, are also threatened. All other rainforest plants and animals depend on them. Their future hangs in the balance too, either through general forest loss or through selective felling, especially for timber. Nobody knows for sure how many tree species grow in the world's rainforests, but there are probably at least 50,000. More than 10 per cent of them are threatened with extinction.

A few valuable tropical timber species have already become so reduced in the wild through international trade that governments have agreed to protect them through the Convention on International Trade in Endangered Species

of Wild Fauna and Flora (CITES). This agreement has been in force for 25 years and is perhaps the most powerful international conservation agreement. Many rainforest organisms, including orchids, spotted cats, rhinos, tigers, primates and parrots, are given a degree of protection by CITES.

The Future of Rainforests

Conserving the world's rainforests is a huge challenge for the 21st century. At present, it seems fairly likely that all unprotected rainforests will be wiped out within the next 50 years unless we, the human race, take urgent action. The disappearance of rainforests undoubtedly means extinctions – not only of charismatic flagship species, but also of the whole supporting web of biodiversity. So what can we do?

A range of initiatives to conserve rainforest fragments has already been devised. These need to be strengthened. At the same time we must develop other, more innovative approaches. One vital point is that conservationists must recognize the part that human beings play in shaping rainforest ecosystems. People have changed every rainforest on Earth and, in an overcrowded world, their impact will no doubt continue to be felt. International, national and local conservation initiatives must recognize this fact and seek ways to minimize our harmful actions.

The best way forward would be to improve the management of rainforests – whether for wilderness conservation, for timber production or for other controlled extraction. We also need to recognize that we must all, in some way, pay for conservation. The conservation of rainforests is too big, and too global, an issue to be left to the poorest people in the poorest countries, already burdened with the legacy of international debt.

LEFT Rainforests are valued for their biodiversity and also for the ecological services that they provide. For some people, the so-called 'existence value' of rainforests is sufficient justification for their conservation.

ABOVE Cloud forests, such as this one in Venezuela, account for about three per cent of tropical forests. They play an immensely important role in protecting the water supplies for lowland areas.

Rainforest conservation: Rio and after

The loss of tropical rainforests became an issue of international concern during the 1970s. Commonly known as the Rio Conference, the United Nations Conference on Environment and Development (UNCED) global summit, held in Rio de Janeiro in 1992, is often used as a benchmark to measure international progress in forest conservation. One of the most important outcomes of the Rio meeting was the Convention on Biological Diversity (CBD). The CBD is an ambitious, wide-ranging treaty with huge potential. It has so far failed to effect real change at an international level, but work continues. It remains important to elaborate the provisions of the Convention and to develop tangible and practical outputs from the various work programmes, including the forest biological diversity programme.

Many conservationists were disappointed when the Rio Conference failed to agree on the development of a new

international treaty specifically dealing with the conservation and wise use of forests. As a compromise, a set of Forest Principles was adopted and after Rio, the Intergovernmental Panel on Forests (IPF), subsequently renamed the Intergovernmental Forum on Forests (IFF), was formed. At its final session recently, the IFF could not agree on a global forest treaty, but did agree that it should remain as a permanent UN forum. This means that government ministers will meet every two years to review existing commitments to conserve forests and manage them sustainably. There is concern from the NGO community (non-governmental organizations) that the forum will just be another 'talking shop', which will lead to no real or tangible action. This reflects a broader sense of frustration with the lack of progress of the various related international forestry conservation initiatives in the years since the Rio Conference.

It is clearly not easy to reach agreement, at an international policy level, on how to save the rainforest. So many issues of economic and political sovereignty are involved. The designation and protection of networks of globally important rainforest sites have been less controversial. A selection of the world's elite rainforest sites is listed under the World Heritage

Convention, an international conservation agreement that has 158 member countries. The Convention is designed to give protection to areas of outstanding cultural and natural heritage. At present, 33 tropical forest sites are listed together with several major temperate rainforest sites. They range from Ujong Kulon National Park, on the Indonesian island of Java, which protects the endangered Javan Rhinoceros, to the Sangay World Heritage site in Ecuador and the Manu National Park in Peru, one of the most species-rich sites on Earth. World Heritage sites deserve to enjoy minimum disturbance as areas for scientific study and global appreciation. Sadly, however, the Convention lists many forest sites as 'in danger'.

Another internationally recognized network of sites operates under the Man and Biosphere Programme. Some Biosphere Reserves in rainforest areas have been designated through this international conservation programme. Biosphere Reserves differ from World Heritage sites in that they specifically allow for research into the interactions between people and the ecosystems where they live. Examples of both World Heritage rainforest sites and rainforest Biosphere Reserves are given in this book.

OPPOSITE LEFT Squirrel monkeys
live in the forests of Central
America, where they feed on fruit
and insects. The rainforest habitat
is essential for the survival of
these animals.

ABOVE Lush streams and rivers
weave through lowland rainforest,
with creepers and low-hanging
branches on all sides.

RIGHT The Hoatzin occurs in
permanently flooded forests of the
Amazon, the Orinoco and the
rivers of Guyana. It is dependent
on certain marsh plants for food,
including the leaves, flowers and
fruits of the White Mangrove tree.
Sometimes, it eats fish and crabs.

Managing the Rainforests

Today, only around two per cent of tropical rainforests have formal protection status under national legislation, sometimes with the added distinction of international recognition. The remaining vast tracts of rainforest that have no direct conservation protection clearly have much to gain from sensitive management. Various international standards of forest management have been developed over the past decade or so.

The International Tropical Timber Organisation (ITTO) was formed in 1986 to help research, promote and co-ordinate trade in tropical timber. The organization set a target that all internationally traded tropical timber should come from sustainably managed sources by the year 2000. They also agreed guidelines and criteria for sustainable forest management.

The Forest Stewardship Council (FSC) also promotes the sustainable management of the world's forests, both tropical and temperate. An independent non-profit organization, the FSC was set up in 1994, partly because it was felt that the ITTO was making too little progress. The FSC's work is based on a set of principles designed to ensure that forests are managed in ways that are environmentally appropriate,

LEFT The Orang-utan is one of the magnificent creatures that will survive only if the destruction of their rainforest habitat in Indonesia is halted.

RIGHT Rainforest soils are usually fragile and nutrient-poor, yet they support lush natural vegetation. If disturbed by human activities, the soils are rapidly eroded.

socially beneficial and economically viable. It acts as an umbrella organization, accrediting and monitoring bodies that certify forests. The products from certified forests can carry the FSC logo, indicating to consumers that the wood or paper goods that they are purchasing have been derived from well-managed forest sites. The FSC has made remarkable progress, but, unfortunately, certification successes have been mainly in the north temperate forest regions. Relatively few tropical hardwoods are available from certified sources.

Improving the management of rainforests for timber production is one way to secure a future for these magnificent and irreplaceable ecosystems. The development of eco-tourism is another, at least in some parts of the world. Appreciating the value of what is being lost through first-

hand experience can only enhance the long-term chances of rainforest conservation.

Governments alone will never save the rainforests. Voluntary organizations around the world have a major role to play. International conservation organizations such as the World Wide Fund for Nature (WWF), Conservation International (CI), and Fauna & Flora International (FFI) all make a significant contribution. Case studies outlining interesting rainforest projects run by these and other conservation groups are included in this book.

All the major conservation groups recognize just how important it is to work with local people, in a variety of ways, to save the rainforest ecosystems. All of us, in our own way, can play a part in ensuring their survival. Saving the rainforests is a global priority.

Chapter Two

Africa

Gorillas in the Mist

The Heart of Africa – dense, majestic rainforest inhabited

by great apes and forest elephants – has long captured

the imagination of intrepid explorers and adventurers.

Major rivers, such as the mighty Congo, have long provided

access to the interior for those lured by the promise of great wealth

from slaves, diamonds and other minerals, timber and ivory.

Intense timber exploitation for European markets began

in West Africa in the Victorian era.

Large swathes of rainforest remained intact until the

close of the 20th century, but in Africa, as elsewhere,

time is now running out.

Rainforests of West Africa

The rainforests of West Africa form a belt running parallel to the coast from Senegal, south east to Gabon. The Dahomey Gap forms a natural break in this rainforest belt, in Togo and Benin, where savanna and dry forest extend to the coast, dividing the West African forest into two major blocks.

Descriptions of Africa's forests often divide the West African forests from those of Central Africa along the boundary between Nigeria and Cameroon. This is an artificial distinction: neither the rainforests nor the species within them recognize national boundaries, of course. The Central African forests extend across a huge area of the continent through Cameroon, Gabon, Congo and the Democratic Republic of Congo (formerly Zaire), forming Africa's equivalent of the Amazon Basin. Farther east, patches of rainforest reach the coast in Kenya and Tanzania.

The once mighty forests of West Africa are now reduced to fragments of their former range. In the Ivory Coast, Guinea, Nigeria and Sierra Leone, the primary rainforests now cover less than 10 per cent of their original extent. Ghana and Liberia have kept more forest: over 20 per cent and roughly 40 per cent, respectively, of their original forest cover.

Most of Liberia's forests have never been surveyed by scientists, and new species remain to be discovered. Liberia is now recognized internationally as the most important country in the region for biodiversity conservation. The rainforests here harbour many species found nowhere else and others that are nearly extinct outside the country. Nearly all the remaining populations of the Pygmy Hippopotamus are found near watercourses and lakes in Liberia's forests. Other important species are the Jentink's Duiker, one of the largest of this widespread genus of forest antelopes, the closely related but smaller Zebra Duiker and the Liberian Mongoose. All these creatures are hunted for meat. Liberia is also home to West Africa's largest populations of African Elephants.

The exploitation of West African forests has long been one of the driving forces in their decline. As long as 120 years ago, Ghana and the Ivory Coast were already exporting

PREVIOUS PAGE Rainforest on Mount Sabinyo Volcano in Uganda, home to the Mountain Gorilla.

LEFT Within Ghana, rainforest is found mainly in the southwest. The local people make use of many different trees, including over 300 species that are harvested for wild fruit.

RIGHT The Pygmy Hippopotamus prefers the shade of deep forest. It is more solitary than the Common Hippopotamus, generally occurring alone or in pairs. Mating usually takes place in water.

valuable hardwoods to Europe. African mahoganies, species of *Entandrophragma* and *Khaya*, were much prized then and, although much depleted, are still in demand. The main commercial species of *Entandrophragma* are Tiama, Sapele and Sipo. As stocks of these timbers have declined in West African forests, the trade has moved to the relatively undisturbed forests of Central Africa. Afrormosia or African Teak is another valuable hardwood of West and Central African rainforests. This timber was little known in world trade before the Second World War, but over the past 50 years logging has severely reduced the species in the wild.

In West Africa as elsewhere, logging has too often been planned for short-term financial gain with no thought for the ecological and longer-term economic impacts. For seven years during the 1990s, a brutal civil war disrupted commercial forestry in Liberia. When the conflict ended, the loggers moved in. In 1999, an Indonesian logging company negotiated a concession of over one million hectares (2.5 million acres) in the southeast of the country. The scene

had been set for Liberia to lose a quarter of its forest areas at a stroke. By February 2000, the company had upgraded or constructed over 150 kilometres (90 miles) of large roads, many of which cut a swathe through pristine forest, and was logging an estimated 60,000 cubic metres (two million cubic feet) of timber per month.

Other major concessions are under negotiation with foreign companies. The only hope for Liberia's forests is that forest management will be radically improved to meet internationally recognized standards. Sustainable management of West Africa's timber resources is a vital key to forest conservation. However, for vast areas of once undisturbed rainforest and its wildlife, it is already too late.

Fortunately, some areas of Liberia's forests are protected for conservation. In 1983, Liberia declared Sapo National Park, its first and only such park, composed of 130,800 hectares (323,200 acres) of lowland rainforest. The park was disrupted during the war years in the 1990s, but still retains important wildlife populations.

Case Study: Fauna & Flora International in Liberia

In a country devastated by war, conservation is unlikely to be a top priority. Looking ahead, poverty and lack of economic opportunity are major problems to be faced by the Liberian Government. Foreign income from logging is clearly attractive; but who benefits?

The Society for the Conservation of Nature in Liberia (SCNL) is run by Liberians who care deeply about the country's forests and wildlife. But faced with huge threats to the forests and wildlife, their conservation efforts need overseas help. At the invitation of SCNL, Fauna & Flora International (FFI) is one organization that has stepped forward to help. In addition to providing small-scale core support and technical assistance to SCNL, FFI raised funds for a survey of the middle Cestos and Senkwehn rivers, an area proposed as a national park. The survey found the area to be of extraordinary conservation value, but threatened by intensive, indiscriminate hunting and spontaneous agricultural settlement following construction of logging roads.

FFI is helping to re-establish management of Sapo National Park and to develop skills and tools for expanding Liberia's protected-area system. The programme, funded initially by the UK Darwin Initiative, as well as WWF, is updating the park's management plan and restarting management on the ground, training Liberians in protected area management planning and field techniques and providing assistance for sanitation and water supply to local communities. Helping local people to reap the benefit from conservation schemes is very important to ensure their long-term success.

Poor information on Liberia's forests has hampered proper management decisions. FFI has therefore teamed up with CI to undertake a reassessment of the entire country's forest cover. Funding from the European Commission is making this work possible. Using satellite imagery, geographic information systems (GIS) and a facilitated process to reclassify protected forests nationwide, the project will establish a detailed database that can routinely be updated. This database will be made available to Government, donors and civil society alike.

Nigeria and Cameroon

The largest country in West Africa, and home to one-fifth of Africa's human population, Nigeria has lost much of its rainforest but has had some conservation successes. The Cross River National Park on the Oban Hills is the country's first rainforest national park. This important protected area is a sanctuary for Nigeria's rainforest wildlife, including 36 globally threatened tree species, seven of which are not known outside the park. Between 150 and 200 Cross River Gorillas live here, and the National Park also helps to protect the habitat of the Drill, a relative of the baboon. One of the most important challenges of the Cross River National Park is to ensure that the needs of local forest-dwellers are met. More than 70,000 people in nearby villages have traditionally used the park for hunting, fishing and gathering forest produce. Ways need to be found to involve these people in park management and encourage them to respect the boundaries of the park.

The southern part of the Cross River National Park joins up with Korup National Park in Cameroon. Korup has been very important in raising public awareness of the plight of African rainforests. In 1982, the first UK television screening of *Korup: an African Rainforest* was widely acclaimed and helped to launch a major campaign to save the area. Korup National Park, designated in 1986, lies close to the Atlantic coast of Cameroon and has an area of 126,000 hectares (311,000 acres). One-quarter of the world's primate species occur here, and half of all the plant species found in Africa's tropical forests.

The importance of elephants

Complex relationships among interdependent species support rainforest ecology. Many relationships are poorly understood. The following examples show the forest elephants' role in the regeneration of three threatened species of African hardwood:

Baillonella toxisperma (Moabi) – Vulnerable. Moabi is restricted to areas of primary evergreen and old secondary rainforest in Cameroon, Congo, Gabon and Nigeria. Overexploited for its timber, it is seriously declining in large parts of its range. It is the second most important exported wood in Gabon. The tree has a variety of other local uses including the production of edible oil which is sold on local markets. Moabi is slow-growing, taking 90-100 years to reach maturity, and regeneration occurs only under a closed forest canopy. Forest elephants are an agent of seed dispersal.

Swartzia fistuloides (Pau rosa) – Endangered. Although widespread in forests in West and Central Africa, this species is rare. It is moderately exploited as a decorative timber known as Pau rosa. There is evidence that regeneration is hampered by the absence of its seed disperser, the African Elephant.

Tieghemella heckelii (Makore or Cherry Mahogany) – Endangered. This important timber species is found mainly in wetter parts of the African rainforest. Overexploitation in some countries is leading to serious population declines, notably in Ghana and in Liberia, where the tree could become extinct. Regeneration may also be limited in parts of its range because of declining numbers of elephants and other seed dispersers.

OPPOSITE LEFT Trees in Rwanda's upland forests are draped with mosses and lichens. Much of the land in this tiny country has been converted to farmland.

BELOW LEFT Imported in great numbers in the past as pets, the Senegal Parrot lives in moist woodland. It is split into three subspecies with different coloured bellies: yellow, red or orange.

BELOW RIGHT This snake is one of the eight arboreal bush vipers of central Africa.

Rainforests of Central Africa

More than 80 per cent of Africa's rainforest is in equatorial Central Africa. Pressures are growing on the forests of this vast area but large tracts remain inaccessible and virtually untouched. Forest creatures here include Pygmy Chimpanzees, Gorillas, Okapis and Elephants, together with a wide range of other primates, birds and a multitude of plants. The biodiversity leader of Africa is here: the Congo, home to more different species of plant and animal than any other African country.

Most of the people living in Central Africa rely largely on the forests for their livelihood. The Bantu and Sudanic people practise shifting cultivation, clearing patches of forest and planting them with crops such as cassava, plantains, corn, and yams. They also grow cash crops such as peanuts, rice, coffee and oil palms. After two or three years the farmers move on, abandoning their forest plots, leaving land that is bereft of trees.

OPPOSITE LEFT Rainforest in the Republic of Congo remains relatively undisturbed, providing a habitat for globally threatened primates such as this young Gorilla.

LEFT In recent years, some groups of pygmies have turned to farming and are now settled in villages. Others have moved to urban areas. This has been encouraged by Government policies and missionaries. Nevertheless, few pygmies lose all ties to the rainforest.

BELOW Nearly all of the Bwindi National Park, Uganda, is still covered by rainforest, including many flower species. It is of outstanding international importance for the conservation of over 90 mammal species that occur there.

The pygmies of Central Africa are still traditional hunter-gatherers. About 200,000 pygmies now live throughout the region, in small groups, with diverse languages and traditions. For several thousand years, pygmies such as the Mbuti of the Congo have coexisted with more settled farmers, whom they supply with forest products. The Mbuti live in the Ituri Forest area, close to the border between the Congo and Uganda. They are skilful hunters of wildlife, travelling through the forest with their bows and arrows in search of prey. Their shelters are dome-shaped huts made from young trees bent and woven together and covered by leaves. They gather caterpillars, beetle grubs and honey to eat, along with a wide variety of plants that they use for food and medicine. They trade wild forest products with local farmers in exchange for cultivated crops, cloth, cooking pots and tools.

The wealth of forest knowledge of the Mbuti and similar tribes is irreplaceable, and could be lost if their ancient ways are allowed to die out.

Threatened species

Although the Central African forests are currently under less pressure than those of West Africa, many species found in them need conservation attention. One such is the Okapi, a close relative of the Giraffe, which is threatened by hunting and habitat loss. This elusive creature may never have been abundant, and was unknown to Europeans until about 100 years ago. It is extinct in Uganda, but a few thousand Okapis live in the eastern Congo, where they have been protected by law since 1933. They are mainly found in the Ituri rainforest, one of the largest remnants of Congo Basin forest and a regional centre of biodiversity. A representative area of this rainforest was designated the Okapi Wildlife Reserve in 1992. Covering 13,726 square kilometres (5,182 square miles), with various types of forest and swamp, the reserve is rich in birds, including the threatened Congo Serpent Eagle and Congo Peafowl. It also has at least 13 species of primate and one of the largest populations of forest elephants in the Congo. The effective conservation of this protected area and its wildlife remains a major challenge. Commercial logging is not yet a significant threat here, but other kinds of pressure on the forest are increased by the immigration of people from other parts of the Congo. There are not enough guards to evict the settlers who clear the forest, hunt game and generally disrupt delicate ecosystems, posing a danger to the forest and its inhabitants in the Ituri region. Illegal gold prospectors have

established numerous camps in the Okapi Wildlife Reserve, apparently with the tacit support of influential people in the region, and there is poaching for ivory. Unfortunately the local Bantu and Mbuti people see the reserve as a restriction on their traditional land-use rights. It will be necessary to involve local communities in park management, and provide basic support for rural development, to ensure the long-term success of this globally important reserve.

Primates are in serious trouble throughout the African rainforests. The development of roads for logging opens up areas that once lay undisturbed, deep within the forest. Roads have also encouraged trade in wildlife sold for food, not only locally but also in urban markets outside the forest areas. Commercial trade in 'bushmeat' is a major threat in logging areas. Traditionally, hunting for food has been essential for the survival of forest-dwelling people, and wildlife is the main source of protein for people who live in the forests. Duikers and similar animals are traditionally the preferred bushmeat, but, as these become scarcer, primates are increasingly exploited too.

One of the most endangered creatures is the Drill, a relative of the baboons. This rather mysterious primate is found only in a small area of rainforest in Cameroon, Cross River State, Nigeria, and the island of Bioko. Little is known about the ecology and habits of Drills, which have scarcely been studied in the wild. It is known, however, that they live mainly on the ground searching for food such as fallen fruit, roots, leaves and insects. They climb trees to pick fruit and to sleep at night. Drills communicate by using a range of facial expressions, vocal sounds and behaviour patterns. Because Drills are noisy and gregarious, hunters can easily locate groups of them within the forest. Dogs are used to track the Drills and hold them at bay while they are shot. Unfortunately, legislation against hunting Drills and other wildlife in Cameroon and Nigeria is largely ineffective.

OPPOSITE TOP The Okapi, a browser on leaves and fruits, is dependent on the rainforest for its survival. An elusive animal, related to the Giraffe, the Okapi lives singly or in small groups.

OPPOSITE BOTTOM The Little Bee-eater is found in most East and Central African countries, where it occurs in a variety of habitats.

BELOW Eco-tourism provides a much-needed source of revenue in the Virunga Volcano region of Rwanda, Uganda and the Congo. Mountain Gorillas are the main attraction.

Case study: The Pandrillus Project

Liza Gadsby and Peter Jenkins, two Americans resident in Nigeria, initiated this conservation project in 1988. One of the first steps was to survey all Drill habitats in Nigeria and Cameroon. This provided information on the number of remaining Drills and the threats they face. During the survey, infant Drills were seen in villages, surviving as orphans when their mothers were shot. In 1991, the Drill Rehabilitation and Breeding Centre (DRBC) was set up, in co-operation with the Nigerian Government, to rescue these animals and raise them into a natural-sized group. The animals were handed over by villagers, wildlife officers or national park staff. In 1996, most Drills were moved to a permanent site on community forest in the Afi River Forest Reserve. Pandrillus is now developing a community-based conservation programme for Afi Mountain where wild Drills, Gorillas, Chimpanzees and other endangered species still survive. In May 2000, the area was designated the Afi Mountain Wildlife Sanctuary. The Pandrillus Project has sponsored a ranger programme employing former hunters to serve as wildlife guides for eco-tourists and to protect Afi Mountain from hunting. As well as tourism, the project has developed an agroforestry tree nursery that grows a range of trees including the valuable native bush mango and ebony species. One of the most important achievements of the Afi Mountain Sanctuary has been in providing an environmental education programme for local people and visitors to the Sanctuary.

Primates at risk

The Pygmy Chimpanzee, Dwarf Chimpanzee or Bonobo is confined in the wild to the forests of the Democratic Republic of Congo. Few studies of Bonobos have been made, but it seems that their populations are very scattered. Recent survey work suggests that Bonobos are common in parts of the Salonga National Park. Created in 1970, this protected area is the largest rainforest park in Africa and probably in the world. In 1984, it was recognized as a World Heritage Site. The Bonobo survives here against the odds, for people in the Salonga region consider its meat to be a delicacy.

LEFT Male silverback Mountain Gorilla, one of the most majestic primates, and most visited, of the Central African rainforest.

ABOVE LEFT Bonobos feed mainly on fruit. They also eat leaves, flowers, fungi, invertebrates and, occasionally, small mammals.

ABOVE RIGHT Mountain Gorilla and baby. Family groups usually

comprise around 15 individuals, led by a mature male.

RIGHT Male Bonobo. In Bonobo society, the basic unit is a mother with her offspring, including mature males. Whereas with Chimpanzees the males are the more sociable, Bonobos form strong bonds between females and between males and females – in other words, there is greater sexual equality.

The Bonobo's larger relative, the Chimpanzee, is more abundant: between 100,000 and 200,000 individuals survive in the wild. Chimps are found in the forests of West and Central Africa, but everywhere they face those all-too-familiar threats: hunting and the destruction of their habitat.

The Gorilla, largest of all primates, has four separate sub-species. The western population is the most abundant, with probably 100,000 individuals, mainly in Gabon. Between 150 and 200 Cross River Gorillas remain in Nigeria and Congo. A population of around 10,000 is found in the eastern Congo, where forest clearance is a major threat. The fourth population, with about 650 individuals, occurs in the Virunga volcano region of Rwanda and the Congo and the Bwindi Forest of Uganda. These are the so-called 'Mountain Gorillas'. Rwanda and neighbouring Burundi are the most densely populated countries in Africa and their mountain forests have largely been cleared.

The Gorilla is a hugely important flagship species for conservation. Through eco-tourism, it provides a source of economic revenue for the poor and war-ravaged countries where it occurs. Gorilla-watching for visitors was the mainstay of the economies of Rwanda and Uganda in the 1980s and 1990s respectively. Few visitors could leave unmoved or uninspired after being lucky enough to see these wonderful but endangered creatures in their natural habitat. The best chance to see mountain Gorillas is in the Volcanoes National Park in Rwanda, where Gorilla-watching tourism was pioneered.

Rainforests of East Africa

Images of dry savannas and magnificent plains wildlife are more usually associated with East Africa than steamy jungles, but rainforests do occur here and are of great biological importance. As well as the montane rainforests of western Uganda, Rwanda, Burundi and western Tanzania, which are an extension of the dense lowland forests of Central Africa, rainforest fragments are also found farther east, in Kenya and eastern Tanzania.

These eastern rainforests are divided into coastal lowland forests and those of the ancient Eastern Arc Mountains, mainly in Tanzania, with remnant outliers on the Taita and Shimba Hills in Kenya. Although highly fragmented, in areas with high rainfall surrounded by dry woodland, these rainforests are botanically very rich, with an extraordinary array of trees found nowhere else in the world. Exciting new botanical discoveries are still being made, even of large trees previously unknown to science. In total, the flora of the Eastern Arc Mountains consists of around 2,000 plant species, over 25 per cent of which are endemic to these unique forests. The forests on the mountains and coastal plain of eastern Tanzania are recognized as one of the world's most important biodiversity hotspots.

The rainforest fragments of East Africa are extremely important for bird conservation. The Sokoke Forest near the Kenyan coast is home to six threatened bird species including the Sokoke Scops Owl and Clarke's Weaver, both of which are endangered. The Sokoke Scops Owl, a small insect-eating owl, was first described in 1965, and between 1,000 and 1,500 pairs are thought to live in the Arabuko-Sokoke Forest Reserve.

Seven threatened bird species occur in the Usambara Mountains in northeastern Tanzania. Two of these – the Usambara Eagle Owl and the Usambara Ground Robin (or Usambara Akalat) – are endemic to the area. Very little is known about the Usambara Eagle Owl, a large owl whose preferred habitat is montane forest, where it probably breeds in holes in old trees. Its diet is thought to consist of squirrels, bats, galagos and insects. Some juvenile Usambara Eagle Owls have been found in areas that have been partially cleared for cardamom cultivation, suggesting that the species may be able to survive in disturbed forest.

The East African rainforests may have far fewer species of mammals than those of West and Central Africa, but they are much richer in species than the drier habitats surrounding them. Only about two per cent of Tanzania is covered by rainforest, yet nearly one-quarter of the country's mammal species are found in the forests. A few mammal species are endemic to these rainforests, such as the little-known Usambara Squirrel.

LEFT African rainforests are home to 17 species of duiker. The physique of these small antelopes – with long necks and short legs – is specially adapted for negotiating the dense forest undergrowth.

RIGHT The distribution of the Usambara Eagle Owl is very restricted, and its forest habitats are under great pressure outside specially protected areas.

ABOVE African Elephants are found within rainforests as well as in savannas and savanna woodlands. Counting elephants in the dense rainforests is difficult, so estimates of the number which survive in the forested countries of Central Africa are unreliable.

No people in East Africa depend entirely on the rainforests for their livelihood, but some forest products are of considerable importance. Wood is used as fuel and building timber, and for poles and dugout canoes. Pit-sawing is usually done by specialists such as the Wahehe people rather than by local villagers. Harvesting of wild foods such as honey and fruits, and of medicinal plants, is much more widespread. Few people hunt wildlife for their subsistence needs.

The Eastern Arc Mountains, such as the Usambaras, Ulugurus and Udzungwas, were uninhabited until Europeans established coffee and tea plantations around 100 years ago. More recently, commercial logging became a major threat. Fortunately, the conservation importance of the Eastern Arc Mountains is now widely recognized though, so far, relatively few areas are managed for the conservation of biodiversity.

The Future for African Forests

It is essential to find alternative sustainable uses for Africa's precious rainforests if they are to survive. Eco-tourism has been an exciting new initiative, undoubtedly offering hope for a select group of sites, but it can never provide the complete answer for the conservation of the vulnerable African forests.

Timber mining is no answer in the long term and nor is the commercialization of bushmeat. Local solutions need to be encouraged and supported. The protection of rainforests is intrinsically linked to the livelihoods of local people: only by ensuring that their needs are met will these unique, complex and fragile ecosystems be saved for future generations.

BELOW Many glorious plants deck the forest canopy and floor in Bwindi National Park, Uganda. This one is carniverous.

RIGHT The rainforest canopy on the slopes of Myhabura Volcano in Rwanda is an ecosystem in itself, with plants and animals adapted to life high in the trees.

Chapter Three

Madagascar
A World Apart

The extraordinary island of Madagascar is unlike anywhere else
on Earth. The world's fourth largest island, it lies in the Indian Ocean,
400 kilometres (250 miles) off the coast of southern Africa. This isolation
has given Madagascar a unique natural and cultural heritage. It is a place
brimming with mysterious creatures, wild, unique forests, secret
coves and dark, hidden backwaters.

The island's rainforests are a precarious paradise. Tomato-red frogs,
brilliantly coloured arboreal chameleons, exotic orchids and wild spices
captivate the senses. Elusive lemurs and stunningly plumaged ground-
rollers find refuge amongst the palm trees and rosewoods of these unique
forests. Spectacularly spined stumped-tailed chameleons snatch their prey
amongst the leaf litter but no poisonous snakes lurk within the
undergrowth – or indeed occur anywhere on Madagascar.

The cool, dense greenery contrasts sharply with the exposed
and eroded red earth of so much of the island.

Myths and ancient traditions help to protect patches of the forest
vegetation and its dependent creatures but modern pressures are rapidly
becoming too great for Madagascar's rampant yet fragile rainforests.

Natural Riches

Madagascar's plants and animals have evolved in relative isolation since the separation of the island from Gondwanaland, the ancient landmass, around 65 million years ago. Biologically this tropical island shows few affinities with either Africa or Asia. It has an amazing diversity of unique species, many concentrated within the remnant patches of rainforest, and is considered to be one of the top global priorities for biodiversity conservation.

Madagascar is roughly the size of France, with an area of about 600,000 square kilometres (232,000 square miles). It includes dramatic escarpments, mountain ranges and a central plateau area, together with gentle foothills and coastal plains. There is great ecosystem diversity, with differing vegetation types according to relief, geology and rainfall. Sadly, around 80 per cent of the original plant cover has been destroyed and much of the terrain is now degraded grassland. In the rainy season the 'great red island' loses so much soil to the rivers and to the surrounding ocean that the red ring around Madagascar can be seen from outer space. Scientists are divided as to whether Madagascar was once completely forested, but generally agree that without major conservation efforts the Madagascan rainforest and its species diversity will be lost within the next 30 years.

A vast diversity

Madagascar is home to a staggering 10,000 or so plant species, 80 per cent of which are endemic to the island. Ten plant families are unique to Madagascar, which is remarkable considering its relatively small size. (Brazil, the richest country in the world for plant species, has no endemic plant families.)

Although Madagascar's animal diversity is less striking, many creatures are unique to the island. The best known are the lemurs: 33 different lemur species are found here, more than half of them in the rainforests. One new species of mouse lemur, *Microcebus ravelobensis*, was not described until 1997. Madagascar has 250 bird species (nearly half of them endemic), 300 reptile species (with 274 endemics) and 178 frog species (the island's only amphibians), all of which are endemic except for several that have been introduced. Most of the frogs are forest-dwelling tree frogs including the beautifully coloured mantellas, which are sought after by international collectors. Madagascar's reptiles include 57 species of chameleon, two-thirds of the world's total, and over 50 endemic geckos. Scientists are still describing many new amphibian and reptile species.

The once mighty rainforests of Madagascar have declined dramatically in extent, but they remain immensely important for species conservation. As well as the animals that depend on the rainforests, as many as 75 per cent of Madagascar's plants grow there too. Dense evergreen rainforest is found in the eastern part of the island, which has a high rainfall of up to 3,500 millimetres (138 inches) per year. A belt of lowland

rainforest occurs below 800 metres (2,600 feet). At higher altitudes, montane and cloud forest is found. In the drier western areas of Madagascar, deciduous forests occur.

Lowland rainforests

The lowland rainforest has a main tree canopy at 25–30 metres (80–100 feet), which is not particularly high in comparison with the rainforests of Africa and Asia. There are relatively few large emergent trees. The diversity of trees within the rainforests is extraordinary, and it is hard to find two individuals of the same species within close proximity to each other. The main canopy typically has trees of the *Euphorbiaceae*, *Lauraceae*, *Rubiaceae* and *Sapindaceae* families. Beneath the main canopy, a lower canopy is made up of small trees and large shrubs. Palms, orchids and ferns are abundant in the lowland rainforest, and bamboos are scattered throughout. Good places to see Madagascar's lowland rainforests are the Mananara National Park (on the coast in the north), and on the protected island of Nosy Mangabe.

Montane forests and cloud forests

The montane forests, growing at altitudes of 800–1,300 metres (2,600–4,300 feet), have a lower main canopy (at 20–25 metres/65–80 feet), with epiphytic plants, as well as lichens and mosses, growing on the branches of trees. More orchids grow in these forests than anywhere else on the island. A good example of this type of forest is within the Montagne d'Ambre National Park. In the north of Madagascar, this area is an isolated rainforest area, surrounded by drier, succulent-rich, lowland vegetation. The montane forest of Montagne d'Ambre has an abundance of tree ferns and many kinds of palms, growing below the taller forest trees. Visitors are attracted by the lush forest, beautiful waterfalls and varied wildlife, including the Crowned Lemur and Sanford's Lemur, which are shy, but quite abundant within the park.

Another area of montane rainforest accessible to visitors is the Perinét-Analamazoatra Reserve. Canopy trees include the Madagascan rosewoods, *Ravensara* trees and species of *Tambourissa*, with, like so many rainforest trees, fruits growing directly from the trunks. Nine species of lemur occur within the reserve, although they are not all easy to see. The forests are amongst the richest in Madagascar for amphibian species.

At higher altitudes, above 1,300 metres (4,300 feet), cloud forests have more stunted trees, growing to around 10 metres (30 feet), with dense tangled masses of moss- and lichen-covered branches.

Man in Madagascar

The island was uninhabited until 1,500 years ago, when the first human settlers arrived by sea from Indonesia. They were the ancestors of the Sakalava people who live in western Madagascar today. These early immigrants practised shifting cultivation, burning the previously untouched natural vegetation. So began the inexorable processes of habitat destruction and species extinctions in this unique Eden.

Later settlers, rice farmers arriving from Indonesia between the 9th and 13th centuries, were the ancestors of the highland Merina people and Betsileo tribes of northern Madagascar. During the past 1,200 years, Africans and Arabs have also settled on the island, and have introduced Zebu cattle. Even today, people in Madagascar measure their wealth by the number of Zebu they own and Zebus now outnumber people here. Many traditional rituals and ceremonies are based on these prized cattle. The animals have

a major impact on the vegetation of the island. Every year, large areas of vegetation are burnt to try to improve the pasture for them. The fires, fanned by the wind, are also helping to wreck Madagascar's last remaining areas of forest. The fires wipe out its special plants and, therefore, the animals and birds, that depend on them.

Madagascar became a French colony in 1895. French settlers put new pressures on the forests by harvesting valuable hardwoods such as ebonies and rosewoods. The endangered tree *Dalbergia delphinensis*, which occurs in lowland rainforest in southeast Madagascar, near Taolanaro, is one of over 30 Madagascan rosewoods that are threatened by selective felling for export or local use. Sadly, the restricted distribution of this species coincides with a proposed site for titanium mining, which threatens all the remaining and unique coastal forest in the area. Other precious minerals lying under the rainforests

LEFT Rainforest of the Perinet-Analamazoatra Nature Reserve. This protected area, with its lush forest vegetation, shy lemurs and rich variety of amphibians, is easily accessible to visitors from the capital, Antananarivo.

ABOVE LEFT Eroded hills southeast of Mahajanga. Deforestation leads to the barren landscapes that are now all too familiar in Madagascar.

ABOVE RIGHT The White-faced Tree Duck is common in Madagascar. It also occurs in Africa and tropical America.

of Madagascar include gold and sapphires. Attempts to extract them seem likely to pose further threats.

Mining is not the only current threat to Madagascar's unique and fragile forests. Wood is cut for fuel and to make charcoal, and Asian logging companies are keen to acquire forest concessions. The main threat is still the local form of shifting cultivation known as *tavy*. Most of Madagascar's population of over 13 million people depend on the land for their livelihood. Forest is cleared by cutting and burning and is replaced mainly by rice cultivation. Other staple crops include maize and cassava. In the familiar rainforest pattern, crops are grown on the nutrient-poor rainforest soils for only a few years, and then the exhausted plots are abandoned. The secondary vegetation that develops is known as *savoka*, and is eventually replaced by grassland as burning takes its toll.

Protecting Madagascar's Heritage

As early as 1921, a French botanist, Perrier de la Bathie, wrote that felling had destroyed 90 per cent of Madagascar's forests. French biologists already appreciated the need to conserve the island's rare plants and animals, and protected areas began to be established in 1927. Even before this, forest conservation measures had been introduced and forest clearance became a punishable offence as early as the 18th century.

Today, protected areas of Madagascar consist of six national parks, eleven strict nature reserves designed to conserve ecosystems and other special reserves created to protect the species that occur there. National parks are open to visitors, whereas the reserves have more limited access. In total, the formal protected-area system covers less than two per cent of Madagascar's land, but in addition a further seven per cent of the land is protected in classified forests and forest reserves.

Madagascar's protected areas have been designated for timber production, but are potentially much more valuable for biodiversity conservation. Carefully controlled eco-tourism offers the greatest potential for generating income for the country's economy while at the same time helping to conserve Madagascar's precious rainforest resource.

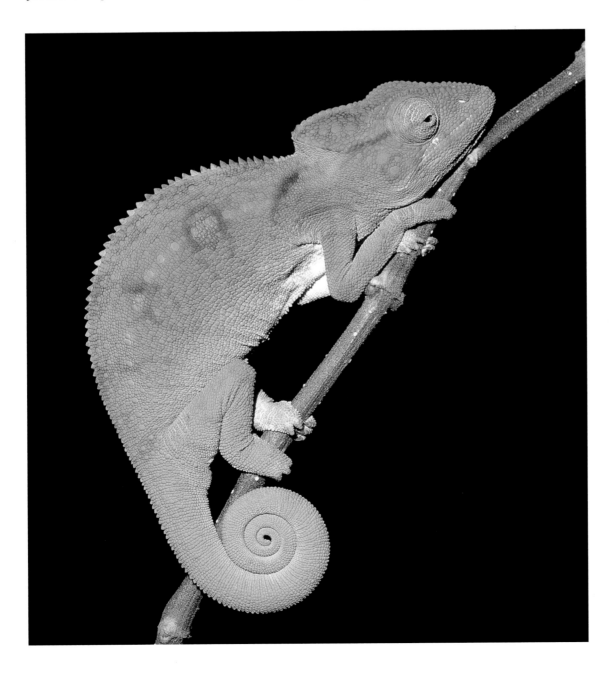

LEFT Oustalet's Chameleon (*Chamaeleo oustaleti*) is the largest of all chameleons, reaching the size of a small cat. This species is still quite common, and is found in protected areas such as the Ankarafantsika Natural Reserve.

OPPOSITE TOP The montane rainforest of Ranomafana National Park, East Madagascar, is an important area for the conservation of globally threatened birds.

OPPOSITE BOTTOM Madagascar Boa (*Sanzinia madagascariensis*). The smallest Malagasy boa, this species lives in rainforest trees in the north and east of the island. It has been collected from the wild for the pet trade.

Case study: Saving Ghost Mountain

Mist-shrouded Ambondrombe, or Ghost Mountain, is situated towards the southern end of Madagascar's rainforest escarpment. The mountain is surrounded by thick forest, which the local people consider to be sacred, so this is one area where rainforest has been conserved.

New migrants moving into the area do not share this respect for the forest and are beginning to clear it for agriculture. Logging is also a growing threat.

The Rainforest Foundation is working with Feedback Madagascar to help local people to save the Ambondrombe forests. One of the most important steps is to make local rice cultivation more sustainable so that the people do not need to clear further rainforest plots. New techniques that are being introduced include the development of composting and vegetable gardening undertaken by the women of the area. A small dam has been built to prevent flooding and to increase the rice crop. The villagers are receiving support to gain legal rights to the forest so that they can prevent loggers from moving in.

The local people are strongly committed to the project and are providing many of the ideas to ensure its success.

Lemurs in Danger

No one has ever made a full evaluation of Madagascar's many threatened rainforest species. The task is daunting. Conservation-status information on some animal and tree species has been assembled, but there is still a long way to go. Of the rainforest lemurs alone, threatened species include the Indri, as well as the Aye-aye, the Hairy-eared Dwarf Lemur, the Golden Bamboo Lemur and the Red-bellied Lemur.

Habitat destruction is a major threat to the Indri, Madagascar's largest lemur. These creatures are now confined to a small area of rainforest where they live in family groups, eating leaves, flowers and fruit during the day and sleeping high in the trees at night. Fortunately, Indris are found on several reserves, including the Zahamena and Betampona Natural Reserves, and the Anjanharibe-Sud and Analámazaotra Special Reserves. This

may not be enough to save the species from extinction, despite the respect suggested by its common local name *babakoto*, meaning the ancestor or father of Man. All lemurs are protected by law, and to some communities it is taboo to kill an Indri, but elsewhere they and other lemurs are hunted as a source of bushmeat and sometimes sold in local markets.

Forest loss is also the main problem for the Aye-aye. Naturally rare and strange-looking, it is considered locally to be an animal of evil omen, so has been persecuted.

The Hairy-eared Dwarf Lemur is one of the least known and rarest of all the lemurs. Until its rediscovery in 1989, it was known only from a handful of museum specimens. It lives in lowland forest in northeast Madagascar, again in areas threatened by general forest destruction.

LEFT The Black Lemur lives only in the northwest coastal forests of Madagascar and the islands of Nosy Bé and Nosy Komba. Males are black and females are golden brown. Newborn babies of both sexes are uniformly black.

ABOVE The Crowned Lemur is protected within the boundaries of the Montagne D'Ambre National Park, North Madagascar. The park also has abundant orchids, palms, reptiles and amphibians, and beautiful scenery, with a crater lake and cascades.

The most remarkable lemur species have long been extinct. At least 15 species have vanished permanently since human beings arrived in Madagascar. These include *Archaeoindris*, which was larger than an adult male Gorilla, *Babakotia* and *Magaladapsis*. Also lost centuries ago were the elephant-birds or giant ostriches, *Aepyornis* and *Mullerornis*. The lemurs of Madagascar are a major eco-tourist attraction and it is hoped that tourist dollars will help protect these precarious creatures.

Plants in Peril

Similar dangers beset the plant life of the Madagascan rainforest. Trees threatened with extinction include the palm *Ravenea louvelii*, which is confined to Andasibe. Fewer than 25 of these palms remain, and rejuvenation is minimal. Another palm species, *Voanioala gerardii*, is in even greater danger, with fewer than 10 examples known to exist in the wild. This palm inhabits primary forest on gentle slopes at about 400 metres (1,300 feet) on the Masoala Peninsula. Deforestation and palm-heart exploitation are the main threats. Unless effective protection can be given to areas of forest on the Masoala Peninsula and the trees are safeguarded against exploitation, the species is unlikely to survive.

Orchids are also in trouble. Madagascar as a whole has one of the richest and most exciting orchid floras in the world.

One in ten of all the island's flowering plants are orchids, and these often exquisite plants are concentrated in the mid-elevation rainforests. Once again the main threat to such plants is habitat destruction. A recently produced catalogue of Madagascar's orchids records many that are known only from plants that botanists collected years ago, in rainforest areas that have subsequently been destroyed. Orchids are also at the mercy of unscrupulous collectors. There are several beautiful species in the genera *Angraecum* and *Aeranthes* which are very popular with orchid growers. Sadly, some wild populations have been plundered to meet the demands for these around the world.

Undoubtedly, more new orchid species remain to be discovered on the island. We must hope it is not too late.

LEFT Flame Tree (*Delonix regia*). This beautiful ornamental tree, one of the most magnificent flowering trees of the tropics, is rare in its native Madagascar, but is now widely planted as a street-tree in other parts of the world. Remaining populations in the wild are threatened by felling for charcoal production. Three other endemic species of *Delonix* are close to extinction.

RIGHT Vanilla Orchid in flower, Ilê Sainte Marie. Vanilla Orchids are cultivated for the beauty of their flowers and for vanilla extract used as a food flavouring.

Chapter Four

India and South East Asia

Forest Treasures of the East

The awesome ruins of Angkor Wat, the largest religious monument

in the world, rise above the Cambodian rainforest. The French

explorer Henri Mohout brought Angkor to the world's attention

in 1860, after it had been lost to civilization for nearly 400 years.

What other treasures, man-made or natural, remain hidden

in Asia's tropical rainforests? With their sheer diversity of forest types,

their unique species and extraordinary landforms, these are

among the richest and most complex forests on Earth.

Hauntingly mysterious even in the 21st century, the Asian

rainforests continue to represent a biological treasure trove

that is scarcely understood.

Asian Rainforests

One of the world's three major rainforest areas is spread across southern Asia, stretching west as far as the semi-evergreen rainforests of the Western Ghats in India and east to the mountainous, tree-clad islands of the Pacific.

The largest of the gaps in this discontinuous stretch of rainforest is in India, where seasonal monsoon forest is the main form of tree cover. In northeast India and from Myanmar (Burma) through Thailand, Cambodia and Laos, to Vietnam and southern China, rainforest and monsoon forest grade into each other. Only fragments of the former richness of these forests now remain, generally in relatively inaccessible highland areas. The character of the different forest areas depends on the amount of rainfall and the level of human disturbance.

True rainforest, with its majestically tall trees and dense crown cover, is the predominant vegetation of Malaysia, Indonesia and the Philippines, particularly in the lowlands. However, few areas of pristine rainforest remain in South East Asia today. The islands of Borneo and New Guinea have the largest undisturbed expanses. About half of Borneo still has rainforest, although much of it has been logged. New Guinea has managed to keep more than three-quarters of its original forest cover. These forests are of immense biodiversity value on a global scale, but their long-term future is far from secure.

People of the rainforests

The rainforests of South East Asia have been inhabited for 40,000 years or more: far longer than those of Africa, where the earliest traces of forest settlement date back a mere 3,000 years at the most. Even in the Amazon, the first people arrived only about 10,000 years ago. The people of Asia's sparsely populated rainforests depended on the rich forest resources for their own use and for trade, but they managed their environment for many centuries in sustainable ways. Precious rainforest commodities such as resins, spices and aromatic oils have been exported from South East Asia for nearly 2,000 years, generally with minimal impact on the environment.

By contrast, the history of rapid rainforest clearance has been short, gathering pace over the past century and speeding up since the Second World War, when timber from South East Asia became a major commodity in international trade. Today an estimated 30 million people live in South East Asia's rainforests and depend directly on the natural resources of those forests. Some indigenous people cling to their increasingly precarious traditional forest lifestyles, but most are forced to accept change. As the forests are lost, so is the indigenous knowledge of the value of biodiversity.

PREVIOUS PAGE Misty morning at Menanggol River, Kinabatangan, Sabah, Borneo.

OPPOSITE LEFT *Rafflesia arnoldii.* This parasitic plant is famed for producing the world's largest flowers. The flowers smell of rotting meat and attract flies, which are thought to be the pollinators.

ABOVE LEFT The Karen people of Myanmar (Burma) are famed for the long necks of women, which result from their extraordinary jewellery.

TOP RIGHT Brightly collared fruits of Creeping Fig (*Ficus aurantiaca*), Gunung Rinjani Forest, Lombok, Indonesia. There are around 800 different species of fig, with diversity concentrated in South East Asia.

BOTTOM RIGHT Cloves are the dried flower buds of a rainforest tree. Native to the Moluccas, commonly known as the Spice Islands, cloves are no longer collected from wild trees. Indonesia is the main producer and kretek, or clove cigarettes, are very popular in the country.

India

Peninsular India's last remaining rainforests are confined to the Western Ghats, the range of hills that extends from the southern tip of India along the west coast to the Gulf of Khambhat, north of Bombay. Rainforest formerly covered large areas of northeastern India, but now survives only in Assam and Nagaland in the foothills of the eastern Himalayas. The Andaman and Nicobar Islands also have areas of surviving rainforest.

The Andaman and Nicobar Islands

The Andaman Islands form a chain south of Myanmar (Burma), roughly parallel to the northern portion of the Malay Peninsula. Their fauna and flora have affinities with those of Myanmar (Burma), whereas the more southerly Nicobar Islands are biologically closer to Indonesia (Great Nicobar Island is only 90 kilometres/56 miles north of Sumatra). The Andamans were known to Marco Polo, who wrote about them and to many other ocean-going explorers: ships have collected fresh water and food here for more than 1,800 years.

The Andamans are home to four groups of indigenous people: the Andamanese, the Onge, the Jarawa and the Sentinelese. On the Nicobar Islands are the Nicobarese and Shompen people. Even today, the Jarawa, Sentinelese, some Onge and Shompen retain their original hunter-gatherer lifestyles. They are expert tree climbers and harvest honey, fruit and nuts. The Onge of Little Andaman traditionally chew the leaves of a special tree called *Tonjoghe* and smear the juice over their arms and legs to prevent bee stings. They also hold the juice in their mouths and blow the fumes to drive the bees away from the hives, enabling them to collect honey safely. Wild boar are a major source of protein and the islanders are also expert fishermen.

OPPOSITE TOP The Indian Flying Fox usually roosts at the same site throughout the year. Flying foxes are very important plant pollinators and seed dispersers.

OPPOSITE BOTTOM The Long-nosed Whip Snake occurs in India. Here it is showing its defensive display.

LEFT The Asian Paradise-flycatcher is commonly found in open forest, woodlands and mangroves. Populations from northern Asia spend their winters in Indonesia and other countries of South East Asia.

Settlement of both groups of islands from mainland India has been a disaster for the indigenous people. The Andamanese population has declined from around 4,800 in the mid 19th century to less than 20 today, mainly through suppression and introduced diseases. The English explorer Portman reported in 1899 that the British Administration was encouraging its officers to suppress the Andamanese by giving them tobacco and alcohol.

The Jarawa have traditionally been more wary of outsiders. The British encouraged hostility between them and the Andamanese and the Japanese killed many Jarawa people during their occupation of the islands in World War Two. The surviving Jarawa retreated deeper into the rainforest, and still live in relative isolation in the western part of the Middle Andamans.

The rainforests of both the Andamans and the Nicobars have remained relatively intact until recent years but are now under severe pressure from agriculture and logging. Andaman Padouk, one of the world's most sought-after tropical hardwoods, is endemic to the Andamans, where it grows in lowland forest.

Mainland rainforests

The Western Ghats of mainland India still have about one-third of their former rainforest cover. The forests at lower altitudes along the coast have been cleared for rice fields and coconut plantations. Tea, coffee, rubber and cinnamon are some of the other crops grown in the hills.

The rainforests that remain are very rich in plant and animal species. More than 4,000 plant species have been recorded in the Western Ghats, about three-quarters of which occur nowhere else.

The rainforests of northeast India are losing ground to a local form of shifting cultivation known as *jhum*. This was a sustainable form of agriculture when the human population was smaller but now it is rapidly degrading India's botanically richest rainforests. *Jhum* is a highly complex form of agriculture involving the production of up to 35 crop species. Pig husbandry, using crop residues and grazing, is traditionally incorporated into the rainforest agriculture. *Jhum* is well suited to local conditions, but, with population pressure, the fallow periods between cultivation have been shortened and the soils and forests are now becoming severely degraded.

Vietnam

Once densely forested, Vietnam had already lost half its forests by the middle of the 20th century. Still more of the tree cover was destroyed during 20 years of war, from 1954 to 1975 and after that there was intensive logging to supply timber for the country's reconstruction. Today less than 20 per cent of Vietnam is forested and these precious remaining forests are still under threat. Conservation efforts have been impressive, but economic development is rapid, bringing with it increasing demands on the forests.

Various types of tropical forest are found in Vietnam. In the north-central part of the country, the natural vegetation below about 1,000 metres (3,300 feet) is tropical lowland evergreen and semi-evergreen rainforest, with tropical montane rainforest above this altitude. The coastal lowlands are almost entirely deforested and in the foothills only fragments remain. Further north is subtropical broadleaf

forest, while in the southern lowlands there are semi-evergreen and montane rainforests, although again much clearance has taken place.

The conservation of the incredibly rich limestone flora and fauna of northern Vietnam is considered a global priority. In wet tropical conditions, limestone ecosystems, including karst and cave systems, are a very special environment with tremendous biodiversity. The forested limestone areas have stunning scenery, such as in Ha Long Bay and locally endemic species are abundant.

Primates and elephants

Vietnam is one of the most important countries in the world for the conservation of primates. Unfortunately, detailed information on the conservation status of particular species remains sparse.

LEFT Ha Long Bay is a popular tourist destination. Cat Ba Island, within the bay, has rugged unspoilt scenery and is a good place to see rainforest. Half of the island is protected as a national park.

ABOVE LEFT The brilliantly coloured Red-shanked Douc Langur lives in the tropical rainforests and monsoon forests of Vietnam, Laos, Cambodia and Hainan. Hunting by human beings for food and habitat loss are the two main threats to this species.

ABOVE RIGHT The White-cheeked Crested Gibbon is an endangered rainforest species. Gibbons are very vocal, well known for their singing. They prefer life in the trees, rarely coming to the forest floor.

Two gibbon species, the Yellow-cheeked Crested Gibbon (*Nomascus gabriellae*) and the White-cheeked Crested Gibbon (*Nomascus leucogenys*), have been listed by The World Conservation Union (IUCN) as Data Deficient, which means that there is not currently enough information to assign a conservation category. The Black-crested Gibbon (*N. concolour*) has been listed as Endangered. At least one species, the Tonkin Snub-nosed Monkey (*Rhinopithecus avunculus*), is close to extinction. The Na Hang Nature Reserve in Tuyen Quang Province is the last known refuge of this attractive creature.

ABOVE The Lesser Slow Loris lives in the forests of Vietnam, Cambodia and Laos. Very little is known about the ecology and behaviour of this appealing species.

OPPOSITE TOP The Lesser Mouse Deer inhabits tropical rainforests and mangroves. This tiny, shy creature is active mainly by night.

OPPOSITE BOTTOM Handsome Red-billed Blue Magpies often fly in succession through open spaces.

Primates have a very important ecological role in the rainforests as agents for seed dispersal and pollination, thus ensuring the continuing survival of rare plant species. Primate populations are under severe pressure from habitat loss due to logging, agriculture, hunting for food and the demand for monkey bones in traditional Vietnamese and Chinese medicine.

International conservation agencies are now helping with various attempts to conserve Vietnam's primates. One example is a project in Phong Nha Nature Reserve in Quang Binh Province which included training forest protection staff, carrying out field surveys and developing a monitoring system for an area that is home to nine primate species. An initial survey was also carried out on Cat Ba Island for the Golden-headed Langur, which is endemic there. Regrettably, recent information on the status of the population is lacking and these primates currently have no special protection.

The Asian Elephants of Vietnam are increasingly endangered, as they are throughout the region. The main causes of the decline of these great animals in Vietnam are the usual ones: habitat fragmentation and loss. Survey results and data collected between 1995 and 1998 show that there are wild elephant populations in 20 areas, with an estimated 109–144 individuals. The elephants are dispersed throughout the country in small, isolated populations of at most 15 individuals. Without active management, Vietnam's wild elephants face extinction.

Case study: Cuc Phuong National Park

Established in 1962 as Vietnam's first National Park, Cuc Phuong is one of the last large fragments of protected forest in the north of the country. The park encompasses 220 square kilometres (85 square miles) of rich tropical forest set on limestone hills and mountains, forming a wonderful green island amid the intensely farmed lowland around it.

Cuc Phuong is ecologically important for its botanical diversity: 28 per cent of all Vietnam's plant species are represented within its boundaries. The park is also home to a diverse variety of endangered and threatened creatures, including Delacour's Langur, Clouded Leopard and Owston's Palm Civet. A total of 88 mammal species is known to occur in the park, but, sadly, some rare and valuable mammals, such as Tigers and White-cheeked Crested Gibbons, that occurred in Cuc Phuong 20 years ago are no longer found there. In addition, over 300 bird species and approximately 50 reptile and amphibian species have been recorded from the park.

The park is valued as an important tourist site, receiving in the region of 40,000 visitors each year. October to January is the peak season; February and March are the best months in which to see flowers and April and May are good for butterflies.

The bones of prehistoric humans have been discovered in several caves in the park, making these sites of national cultural importance. Additionally, local communities retain strong ties to former village and grave sites within the park.

Some 2,000 residents, mostly of the Muong ethnic minority, live in the park, with 50,000 more living in the surrounding buffer zone. Many of these people depend on the park's natural resources for their livelihood, but the forest ecosystem is suffering as they cut timber, collect firewood and other forest products, clear land for agriculture and hunt many of the wild animals.

The Cuc Phuong Conservation Project was established by Fauna & Flora International (FFI) to tackle these threats from neighbouring people. The project has involved working with local communities to understand their needs and promoting conservation by, for example, establishing baseline information about the park's fauna and flora and setting up conservation clubs in local schools. There is also an education programme for visitors and aspects of the park's management system are being improved. It is hoped that these measures, together with increased protection efforts, may begin to address the park's conservation problems and help give Vietnam's fragile wildlife a more secure future.

Cambodia

The great historic kingdoms of Cambodia cleared most of the lowland rainforests in the central areas of the country, but the uplands were left largely untouched until the terrible conflicts of the 20th century. When warfare began in the 1970s, defoliation and bombing were particularly destructive in the eastern mountains. Ironically, Cambodia's years of bitter bloodshed have helped to protect much of the country's wildlife, keeping large areas of habitat inaccessible because of the threat posed by the Khmer Rouge and by the millions of land mines left during three decades of strife.

Recently, Cambodia has seen a significant increase in political stability and security and refugees who fled the country are now trickling back. As the situation improves, however, new pressures are placed on Cambodia's natural resources. Illegal logging has rapidly become a serious concern. The illegal trade in endangered wildlife is also extensive, as in neighbouring countries. In the midst of the post-war quest for economic recovery, there is an urgent need for increased environmental awareness and the development of conservation programmes. A recent report by the Asian Development Bank has predicted that unless urgent action is taken, all Cambodia's forests will be lost by the year 2010.

Case study: The Cardamom Mountains

Swathes of natural rainforest still cloak the Elephant and Cardamom Mountains. The Cardamom Mountains Region spans more than one million hectares (2.5 million acres) and comprises three main massifs: Mount Samkos, the Central Cardamom Mountains and Mount Aural. Ranging up to 1,771 metres (5,810 feet), the mountains are covered with a variety of natural forest types according to altitude, aspect, geology and hydrology: dry deciduous forest, semi-deciduous

forest, lowland and hill rainforest, bamboo thickets and pine
forest. Human density is very low and settlements are almost
wholly confined to lowland areas near rivers. Part of the
range is protected, at least on paper: the Mount Samkos
Wildlife Sanctuary (3,338 square kilometres/1,288 square
miles) to the east and the Mount Aural Wildlife Sanctuary
(2,536 square kilometres/979 square miles) to the west.
Between the sanctuaries, the Central Cardamom Mountains
have been parcelled into logging concessions.

Like many forest areas in Cambodia, both the Cardamom
Mountains and remote areas of Mondulkiri Province
provided a refuge for the Khmer Rouge, who forced many
local villagers to join their cause or flee to refugee camps in
neighbouring Thailand. Thus, while valuable biological
information has been gathered from aerial and satellite images
and interviews with hunters, very little biological fieldwork
has been done in these areas until recently. Many areas and
most taxa have remained unknown. This has hampered efforts
to develop effective strategies for conservation.

Large mammals and their habitats were the primary focus of
surveys undertaken in 1999–2000 in order to determine which
species and habitats needed protection most urgently. A wide
range of other taxa – including plants, bats and other small
mammals, reptiles, amphibians and insects – were also surveyed
and their conservation needs assessed. Ten foreign biologists,
specializing in a wide range of South East Asian taxa, undertook
the survey working with local counterparts from the Wildlife
Protection Office and the Ministry of Environment.

The survey work has produced stunning results and
confirmed that the Cardamom Mountains are one of Asia's
richest biodiversity areas. As a result of the field surveys and
resulting publicity, Cambodia has pledged to lift the threat of
logging from the Cardamom Mountains.

OPPOSITE LEFT The mystical
Mount Khmaoch or Ghost
Mountain abounds in local
legends. Clothed in evergreen
forest, this area provides a
sanctuary for elephants.

ABOVE TOP Endemic to the
Cardomom Mountains, the
Cardamom Banded Gecko is
found only in pristine rainforest.

ABOVE BOTTOM *Rafflesia
himalayana* in Mount Samkos
Wildlife Sanctuary – part of an
important biodiversity-rich area
that is being proposed as a World
Heritage site.

Laos

The landlocked country of Laos, sandwiched between China, Cambodia, Thailand and Vietnam, is one of the poorest countries on Earth, its economy being heavily dependent on its neighbours.

The majestic Mekong River flows through the rugged mountains of Laos on its way to Cambodia, Vietnam and the South China Sea. Three-quarters of the population of Laos live along the Mekong and its tributaries. Fragments of rainforest are found in the central mountain areas and also in small patches on the Mekong River Plain, but about half the rainforests of Laos have been destroyed or degraded. Shifting cultivation has been mainly to blame and uncontrolled logging has recently become a major threat.

LEFT Kuang Si Waterfall, Luang Prabang area, Laos.

ABOVE Sunset on the Mekong River. The Mekong is the principal river of Laos and is the main route for transportation within the country.

RIGHT The Coral-billed Ground-cuckoo, with its conspicuous red bill and legs, is a very shy bird. It lives in dense primary and disturbed rainforest.

Thailand

A mere 50 years ago, great tracts of monsoon forest and rainforest covered about two-thirds of Thailand. Today, rainforest cover has dwindled to around 15 per cent of the land and is found only as fragments in south peninsular Thailand. Monsoon forest covers a further six per cent of the country. Lowland rainforests have been heavily logged and opened for extensive economic development, while peat swamp forests, which occur in tiny areas, have been threatened mainly by agricultural development. Wildlife populations are under threat from illegal hunting.

Historically, Thailand has relied on the export of hardwoods, in particular Teak, as a major source of foreign exchange. Foreign trading countries started commercial exploitation of Thailand's northern Teak forests in the middle of the 19th century. Teak remained the most important commercial timber until after the Second World War when international trade in dipterocarps from the lowland rainforests began to develop. In 1889, ownership and control of all forests was transferred to the Thai Government and laws were introduced to prevent the excessive exploitation of Teak forests.

OPPOSITE TOP Bamboo has a multitude of uses in tropical countries. Here, a bamboo pot is being used to carry water.

OPPOSITE BOTTOM Yao hill tribal mother and child in northwest Thailand. About 75 per cent of the people of Thailand live in rural areas. Rice is the main cultivated crop. Shifting cultivation continues to be one of the main agents of deforestation.

LEFT This fig tree in Khao Yai National Park, Thailand, shows the buttress roots that are typical of tropical rainforest trees. The vegetation of Khao Yai is mainly lush rainforest. Mammals that occur there include Asian Elephant, Sun Bear, Leopard and gibbons.

BELOW The Spectacled Spiderhunter pictured here on a wild banana flower, is found in Thailand, Malaysia and Indonesia. The cup-shaped nests of spiderhunters are made from leafy material sewn together with cobwebs. Both males and females incubate the eggs.

Commercial logging has contributed significantly to deforestation in Thailand and has opened up areas to shifting cultivation and illegal logging. Following landslips resulting from devastating floods in 1988, the Thai Government acted to end commercial logging throughout the country, 'to protect the public interest and prevent natural disasters'. As a result of two decrees passed by the King of Thailand in January 1989, over 300 logging concessions were cancelled and 68 concessions that overlapped protected areas were also revoked.

Even before the logging ban, Thailand was a net importer of timber. The true extent of the import trade has been difficult to assess because of timber smuggling across the land borders with Myanmar (Burma) and Laos. Despite the logging ban of 1989, illegal logging remains a serious problem, particularly in the north of the country. Many species are affected including the rainforest dipterocarps and species of *Pterocarpus*.

Thailand has a good network of protected areas, with various different categories of land designated for nature conservation. More than 50 national parks protect different types of forest, but still, especially in lowland areas, the protected area network includes too little rainforest.

Malaysia

The natural heritage of Malaysia includes some of the world's most magnificent rainforests. Nearly half of the total land area of Peninsular Malaysia is clothed in rainforest. The Malaysian states of Borneo are still relatively well covered too: Sabah's rainforests cover around 40 per cent of the land, while the total forest area of Sarawak, at around nine million hectares (22 million acres), represents well over half of the total land area of the state. The main forest types in Malaysia are lowland and hill dipterocarp forest. Peat swamp forest is less extensive everywhere; most has been cleared for cultivation in Peninsular Malaysia, but Sarawak retains about 11 per cent. Sabah still has this forest type in small coastal areas.

Much of the lowland forest of Peninsular Malaysia has been lost to rubber and oil-palm plantations, to agriculture and to urban development. These forests have been cleared as part of Government land-use policy. The Permanent Forest Estate incorporates all the remaining forestland that is retained for timber production and forest conservation. Increasingly the hill forests are used for timber, since the lowland forests have been logged and then converted to agriculture.

Timber is one of Malaysia's main exports, in the form of logs, sawn timber or manufactured products. The Malaysian dipterocarp forests have at least 3,000 timber species, of which over 400 have been traded internationally. There are about 142 commercial-timber tree species in Sarawak and Sabah trades in around 30 timber species.

The forests of Peninsular Malaysia have been managed for timber production for around 100 years and logging here is relatively well managed compared with that in other parts of the world. Initially the timber was mainly for local use, with the export industry developing to supply timber needs especially in the United Kingdom. Harvesting was on a relatively small scale until the 1950s, when international demand for Malaysian timber became significant and exploitation became more intensive, providing an important source of export revenue. International demand increased rapidly after the end of the Korean War in 1953 and agricultural schemes implemented after Malaysia's independence in 1957 made large quantities of logs available. The development of an export market for timber from Sabah and Sarawak was well established by the early 1970s and Sarawak's timber exploitation increased dramatically during the same decade. By 1976, Sabah accounted for nearly half of Malaysia's total log production.

Logging followed by agricultural conversion has accounted for nearly half of Sabah's deforestation over the past 25 years. Shifting cultivation has been the other main cause of forest loss. Sarawak's deforestation has likewise been the result of shifting cultivation and logging for the export market. The peat swamp forests were among the first areas to be logged in Sarawak and by 1979 almost all these forests had been licensed for timber extraction.

LEFT Rising clouds in Taman Negara, one of Malaysia's magnificent national parks.

ABOVE The spectacular shapes and colours that occur in the rainforest are often more eye-catching than anything seen in an art gallery.

Traditional lifestyles

Malaysia is a rapidly modernizing country but many of the people still rely directly on the rainforest for most of their needs. The population of Sarawak is ethnically very diverse. Indigenous people include the Kajang, the Kayan, the Kelabit, the Kenyah and the Penan. The traditional way of life for such people is focused on longhouses, which can be home to up to 100 people. Constructed from local hardwoods, the houses are raised on stilts to give protection from flooding and wild animals. Ladders and trapdoors are the means of entry. Inside the longhouse, mats and baskets are woven from rattans. Feathers, skins, antlers and tusks are used for decoration and traditional rituals. Until quite recently, feuding was common and headhunting was a traditional activity.

For thousands of years, hunting has been an important part of the culture of the forest-dwelling people of Sarawak and elsewhere in Borneo. Archaeological evidence of hunting has been recorded from 40,000 years ago. Nowadays, wild animals are usually caught or killed in Sarawak by means of traps, blowpipes, spears and guns. Blowpipes were traditionally made from hardwoods such as Belian and the darts from palm material. Nowadays, blowpipes are very rarely used for hunting, except by the Penan people.

The Bearded Pig is a favoured animal hunted for meat throughout Borneo. The diet of the omnivorous Bearded Pig includes small animals, carrion, insects, roots and rotting wood. Fallen fruit, including the *engkabang* nuts from dipterocarp trees, is also consumed. During the irregular fruiting seasons, large herds of pigs move through the rainforest in search of fruit. Many hundreds of Bearded Pigs are killed by hunters during these periods of migration.

Species at risk

The Malayan Tapir is found in forests in all states of Peninsular Malaysia together with Sumatra and in southern parts of Myanmar (Burma), Thailand, Cambodia and Vietnam. It is a primitive mammal, thought to have changed little in millions of years. Essentially nocturnal, this tapir is dark coloured at the front and rear, with a broad white band around the middle of the body. This distinctive coloration serves as camouflage in the light and shadows of the rainforest habitat. The main threat to the Malayan Tapir throughout its range is forest loss and degradation. It is quite selective in its feeding habits, usually eating only the young leaves and twigs of a limited range of trees and shrubs. Tapirs do not seem to survive well in disturbed forest. Fortunately, this vulnerable species is legally protected in Peninsular Malaysia and law enforcement appears to be effective.

The world's smallest bear is also found in the forests of Malaysia, as well as in Indonesia, Vietnam, Laos, Myanmar (Burma), Thailand and southern China. The Sun Bear grows to a height of around 1.2 metres (4 feet). It eats fruit, flowers, eggs, honey, birds, small mammals, termites and other insects. Little is known about the size of Sun Bear populations, but they are thought to be declining and the species is considered to be vulnerable. Forest loss is the main problem, but the bears are also killed for food and 'medicinal' use.

Many of Malaysia's plants are also in danger, including over 700 globally threatened tree species. Loss of habitat is again the main reason, particularly for many of the trees and other plants that are naturally rare in the rainforests. Many valuable plants are also threatened by overexploitation. Nearly all the 104 rattan species that occur only in Peninsular Malaysia are threatened with extinction and few are established in cultivation. It is uncertain how many of these 104 species occur within Taman Negara (see the next section), but the illegal removal of commercial species remains a threat within the protected area.

Orchids and pitcher-plants are two attractive groups that have been targeted by unscrupulous plant collectors for the world market in exotic plants. One of the world's rarest orchids, *Paphiopedilum rothschildianum* is an example of a beautiful plant that could easily be lost for ever from its natural habitat. This spectacular slipper orchid is one of 1,200 orchid species found in the forests of the Mount Kinabalu National Park in Sabah, where it grows on rocks. It was cultivated by orchid enthusiasts in the time of the Victorian orchid craze and then, for nearly 70 years, was believed to be extinct in the wild. Rediscovered on the lower slopes of Mount Kinabalu in 1959, it was found 20 years later in a second locality. Despite extensive field surveys, no other sites for this very rare orchid have been found and the area where it grows is threatened by logging, mining and shifting cultivation. Sadly, although the orchid can be propagated in nurseries, theft from the wild remains a constant threat.

The Giant Malaysian Pitcher–plant also grows only in the Mount Kinabalu National Park. Four localities are known for this species and it is thought that several thousand plants remain. It is a spectacular insectivorous plant, catching its insect prey in enormous pitchers up to 35 centimetres (14 inches) long and with a capacity of up to 2 litres (3½ pints). Though not an easy plant to grow, it is eagerly sought after by collectors of insectivorous plants. Fortunately it is protected by law, as are all the species within the Kinabalu National Park.

LEFT Of the eight subspecies of Tiger, the Bengal Tiger is the most abundant. Nevertheless, only about 4,000 remain in the wild in India, Bangladesh, Myanmar (Burma), Nepal and Bhutan.

BELOW Pitcher plants (*Nepenthes ampullaria*), Malay peninsula. These vividly-coloured carnivorous plants extract nutrients from dead insects to supplement the poor nutrient supply of mountain forest soil. They are mostly creepers that climb trees and shrubs.

Protected areas

Malaysia's protected areas include national parks, wildlife reserves, wildlife and bird sanctuaries, state parks, nature monuments and designated protected forests such as the Virgin Jungle Reserves.

Peninsular Malaysia has two national parks. The first, Taman Negara, an area of species-rich lowland rainforest, was established in 1938 as the country's first national park. Covering more than 4,300 square kilometres (1,660 square miles), it is in the central part of Peninsular Malaysia, in the states of Pahang, Kelantan and Terengganu. Its varying forest habitats are home to a vast diversity of plants, including at least 20 important wild fruit tree species and all the large mammals of Peninsular Malaysia including the Sumatran Rhinoceros. More than 300 bird species have been recorded in the park. Nearly all of these depend on forest habitats for their survival.

Endau Rompin National Park, established in 1989, is located in the southeast of Peninsular Malaysia on the borders

LEFT Dipterocarp rainforest, Taman Negara National Park. It is estimated that one-third of the flora of Peninsular Malaysia grows within the park and 22 tree species are known only from the park.

ABOVE Sumatran Rhinoceros – a critically endangered species, but protected within Peninsular Malaysia's national parks.

of Johor and Pahang States. The rainforests of the park provide another relatively safe haven for Peninsular Malaysia's remnant populations of the Sumatran Rhinoceros. Outside the two national parks, very few of these vulnerable creatures remain and even within Endau Rompin visitors would be outstandingly lucky to catch sight of one. Other big mammals here include Asian Elephants, Tigers, Leopards and Sambar Deer. More than 200 bird species have been recorded within the park and many unique plants are found only in this area.

Sarawak has 10 national parks, three wildlife sanctuaries and several nature reserves. The largest protected area is the Lanjak-Entimau Wildlife Sanctuary, with 1,690 square kilometres (652 square miles) of hill forest. This area was designated in 1983,

specifically to protect the Orang-utan. The sanctuary is closed to tourists, but the adjacent Batang Ai National Park has accommodation and visitor facilities. Orang-utans also live there, although they are generally difficult to see.

Other primates enjoying the relative safety of Lanjak-Entimau's protected forests include the Bornean Gibbon and the White-fronted Leaf Monkey. More than 200 different species of bird here include seven hornbill species and there are probably 150 reptile and amphibian species. Lanjak-Entimau and Batang Ai National Park join up with Betuang Karimun National Park across the border in Kalimantan to form one of the world's most important protected rainforest areas.

In Sabah, nearly 20 per cent of the forest is protected. There are six national parks, together with wildlife reserves managed by the State Forestry Department. Sabah's largest protected area is the Crocker Range National Park, covering 1,400 square kilometres (540 square miles) of forest. As yet, this area is little visited, in contrast to the better known Mount

Kinabalu National Park, which attracts around 200,000 visitors each year. Established in 1964, Mount Kinabalu's protected area covers nearly 780 square kilometres (300 square miles) of diverse forest and montane habitats. South East Asia's highest mountain, Gunung Kinabalu, is the crowning glory here. One in 10 of all the park's visitors climbs to the top.

The biodiversity richness of Kinabalu is extraordinary. An estimated 5,000 plant species grow within the park boundaries, a flora more than three times as rich as that of Britain. Many plants are extremely rare and little known, with over one-third documented from just a single locality. The area also boasts an amazing total of 289 bird species, a similar number of butterflies and around 100 kinds of reptile.

Kinabalu is undoubtedly one of the world's botanical treasures, but it is not without conservation problems. Plant thieves are just one of the threats to the park's biodiversity. In 1970, an area in the southeast of the park was lost to the

Mamut copper mine. The park acquired a larger area two years later, but the vulnerability of its boundaries continued. In 1984, another piece of land in the southern part of the park, containing 35 species of oak, was removed from the park by a change in legal status and the forest converted to a golf course, a dairy farm and vegetable plots. 'Compensation' took the form of additional logged forest to the east of the park. To ensure the long-term conservation of Kinabalu, more resources and staff are needed and the communities living around the park need to benefit from its international recognition.

Other sites important for the conservation of rainforests are the Virgin Jungle Reserves (VJRs). A VJR network was begun in Peninsular Malaysia in 1950. Although they do not have legal status, VJRs are established in areas of forest exploitation as permanent nature reserves, natural arboreta or research sites. At present, there are about 80 VJRs, covering 91,000 hectares (225,000 acres), located in various forest types. Unfortunately, logging has been carried out in most of them. Originally it was intended that each VJR should be buffered by managed forest, but the buffer forest has now disappeared from various sites. Pressures on the vulnerable wildlife of such reserves can only increase.

Chapter Five

Indonesia and the Philippines

Islands of Glory

Images of huge palls of smoke and the plight of orphaned Orang-utans were poignant reminders of the fragility of the vast South East Asian rainforests when widespread forest fires hit the international press several years ago. The pace of the destruction of the rainforests of Indonesia and the Philippines has been truly alarming. Yet within the towering rainforests of this region there are still areas of extraordinary natural beauty, populations of some of the world's most awe-inspiring plants and animals and timeless human lifestyles scarcely influenced by the modern world.

Indonesia

With nearly 10 per cent of the world total, Indonesia is second only to Brazil in the area of tropical rainforest it has managed to keep. Tropical rainforest is the dominant climax vegetation of the country and is the most extensive forest type here. In seasonally dry areas, particularly in the southern and eastern islands of the Moluccas, such as Halmahera, Obi, Buru and the Banda Sea islands, tropical rainforest is replaced by monsoon forest. Mangrove, coastal, swamp and peat-swamp forests are different types of rainforest found within this huge and diverse country.

Biogeographically, the forests of Indonesia can be divided into three groups. The Asian region, comprising Sumatra and Kalimantan, is dominated by the tree family *Dipterocarpaceae*. The Australian region, which includes Irian Jaya, the Moluccas and Nusa Tenggara, is dominated by *Araucariaceae* and *Myrtaceae*. A third, transitional region, incorporating Sulawesi and Java, is dominated by *Araucariaceae*, *Myrtaceae* and *Verbenaceae*. Botanically the richest areas are the primary lowland rainforests of Kalimantan and Irian Jaya.

The island of Sumatra is the second largest of the Indonesian archipelago. Large areas of Sumatra's natural tree cover – mainly tropical evergreen rainforest dominated by dipterocarps – have been logged or cleared for agriculture and industrial plantation. Much of the remaining forest is on the uplands that form the western backbone of the island. Heath forests grow in the east of Sumatra and lowland peat swamp and mangroves along the eastern coasts. Mountain areas in the north support natural forests of the pine *Pinus merkusii*.

The Mentawai Islands: Effects of isolation

The Mentawai group of islands to the west of Sumatra are still verdant with tropical rainforest. The islands have been separated from Sumatra for about one million years and this long period of isolation has encouraged many endemic plants and animals to evolve here. The flora is still poorly known, but around 15 per cent of the plants that have been studied are endemic, as are about 65 per cent of the mammals. These include four species of primate: Kloss's Gibbon, the most primitive of all gibbon species, the Mentawai Leaf Monkey; the Mentawai Macaque and the Pig-tailed Langur, a strange creature usually placed in a genus of its own. All these primates depend on the rainforest to some extent. Three of the species favour rattan fruits, which are found only in mature primary forest.

The tropical rainforests and the wildlife of the Mentawai islands are increasingly under threat. Siberut, one of the islands in the group, was designated a Biosphere Reserve in 1981. The

PREVIOUS PAGE Indonesian rainforest overlooking Alas and Ketambe Rivers, Gunung Leuser, Sumatra. The Gunung Leuser National Park is thought to be the most species-rich of all of Indonesia's national parks.

LEFT Facing an uncertain future, the conservation of the Sumatran Tiger depends on protection of the forest resource. At present, logging and forest fires appear to run out of control.

RIGHT Man canoeing on Gangsal River, Bukit Tigapuluh National Park, Sumatra. Even today, Indonesia's rainforests support many people living a traditional way of life.

western part of the island is protected as a national park, but, during the year 2000, the eastern half was divided up for logging. Approval has been given for a large oil-palm plantation in the buffer zone of the national park and there are plans to settle over 10,000 transmigrant families to harvest oil palm from new plantations on the island. The indigenous Siberut islanders, who live a traditional lifestyle, are not content with the changes that rainforest loss will inevitably bring.

The traditional lifestyle of the Siberut islanders had much in common with that of indigenous people throughout Indonesia. The islanders are, however, less dependent on cultivation than are other Indonesian islanders. Crops of sago and taro, which grow in natural swamps, continue to supply the staple food, supplemented by protein from fishing, raising pigs and hunting primates. The traditional economy was ecologically sustainable, with cultural and religious controls on the exploitation of forest resources. For example, local custom dictates that the Kloss Gibbon may be eaten only by young

men, may not be hunted while it is calling, nor be taken into the *uma* or clan dwelling. Clan members share food freely and traditionally there has been no need to acquire wealth. Until the beginning of the 20th century, Siberut's main contact with the outside world was through traders from Sri Lanka, Malaysia and Sumatra who exchanged metal implements, cloth and beads for rattan, copra and timber.

The indigenous people of Siberut have unfortunately scarcely benefited from some 30 years of small-scale commercial logging on the island. Very few are employed by the logging companies and those who are work only as long as they need to earn a desired amount of money. The logging companies have generally ignored traditional rights of ownership to land and to trees, including the highly valued Durian trees. Traditionally, the Siberut islanders considered that felling trees was wrong without first holding a ceremony to apologize to the souls living in the forest for any damage that might be done to their homes.

Rainforest distribution elsewhere

Intensive cultivation has wiped out virtually all the lowland forests in Java, one of the world's most densely populated islands. Small remnant patches of evergreen rainforest have survived on south-facing mountain slopes and monsoon forests occur in the centre and east of the island. Teak plantations have taken over areas unsuitable for cultivation.

Further east, the rainforests of the Lesser Sunda Islands, also known as Nusa Tenggara, are less lush than those elsewhere in Indonesia. Burning has destroyed large swathes of forest and savanna woodland is now the main type of vegetation in most places. Evergreen rainforest survives in isolated patches in steep valleys and the island of Timor has good natural Sandalwood forests.

Kalimantan, the Indonesian portion of Borneo, has the largest expanse of tropical rainforest in South East Asia. Lowland rainforest grows on land up to 1,300 metres (4,300 feet), though logging has had a heavy impact on the lowland

LEFT The Green Peafowl was once common in forests from northeast India to Indonesia. Forest loss and hunting have taken their toll on this beautiful bird that is now rare or locally extinct throughout its range. Fortunately, it is protected in Ujong Kulon and several other reserves in Java.

OPPOSITE TOP LEFT Lesser Bird of Paradise (*Paradisaea minor*) Irian Jaya. This spectacular bird of paradise displays in groups with other males with inviting shakes of their plumes and staccato notes. The female then chooses a mate.

OPPOSITE TOP RIGHT King Bird of Paradise (*Cicinnurus regius*) Irian Jaya. The male of this species courts females by puffing out his plumage whilst uttering repeated screams. It is otherwise inconspicuous, hiding in lowland rainforest.

OPPOSITE BOTTOM A Black Swallowtail Butterfly feeding on wet sand along the Ella Ullu River, West Kalimantan. Often male swallowtails congregate to feed at river margins or other areas that are rich in nutrients. Some species are attracted to animal dung or urine.

forests. There are also extensive dipterocarp forests on the hills and various montane forest formations. In Kalimantan, the only lowland forests that have not been logged are those in protected areas.

Sulawesi, east of Kalimantan, has extensive tracts of primary hill and montane tropical rainforest, with few dipterocarps. Large areas have been cleared for agriculture. Sulawesi's fauna is particularly fascinating, with many endemics. Nearly all the mammals, except for the bats, are endemic to the island and 27 per cent of the birds are found nowhere else in the world.

The largest islands of the Moluccas are Halmahera and Seram, both of which have tropical rainforest and rich montane forests. Irian Jaya, the Indonesian portion of New Guinea, has large tracts of primary tropical evergreen rainforest, dry evergreen forests in the monsoonal southeast and montane forests above 1,000 metres (3,300 feet). There are also extensive mangrove forests here.

LEFT *Balanophora* species. This parasitic plant has a foxy smell that attracts fly pollinators. There are about 16 species in this genus.

RIGHT Rainforest tree with mosses and lianas, Gunung Leuser. Each rainforest tree is a mini ecosystem in its own right, providing a habitat for an array of epiphytic plants and associated invertebrates. Indonesia has over 550 globally threatened tree species, many of which are not yet protected.

Forest loss

The rate of destruction of Indonesia's ancient rainforests is really alarming. In the past 10 years alone, 15 million hectares (37 million acres) of Indonesian forest have been logged and an additional five to 10 million hectares (12–24 million acres) have been destroyed by fire. Rates of deforestation are second only to those of Brazil. The current annual rate of deforestation is estimated at 20,000 square kilometres (7,700 square miles), with logging contributing to about 20 per cent of this forest loss.

Lowland rainforest, swamp forest and mangrove forest are all important sources of timber in Indonesia. Timber is very important to the Indonesian economy and is the second largest export after oil and natural gas. Large-scale exploitation of the rainforests for timber began in the 1960s and the lowland rainforests of Sumatra and Kalimantan have subsequently been particularly heavily logged. In 1980, Indonesia introduced policies to reduce the export of logs so that more timber-processing could take place within the country. The export of raw logs has been banned since 1985.

The Indonesian timber market is dominated by the *Dipterocarpaceae* (species of *Dipterocarpus* and *Shorea*), grouped according to strength as 'red or white *meranti*' and species of *Dryobalanops*, sold as *kapur*. Until about 10 years ago, dipterocarps dominated commercial timber production in Indonesia, but a wider range of species is now being extracted from the forests. One of the main consequences of selective logging, which entails the removal of up to 20 trees per hectare (eight trees per acre) in East Kalimantan, is the loss and degradation of genetic resources of the primary forest. 'Creaming' or 'high-grading' the best individuals of commercial species (*Shorea*, *Dipterocarpus* and *Dryobalanops*) leaves only smaller and genetically inferior trees in the residual stands to provide seeds for the next timber harvest.

The vulnerable Sulawesi Ebony (*Diospyros celebica*) is one of several high-value Indonesian timber species that have long been traditionally exploited and are now of grave concern to local and international conservationists. Others are *Kalappia*

excelsa, *Pericopsis mooniana* and *Santalum album*. Sulawesi Ebony grows only in lowland rainforest in central and northern Sulawesi. It has been heavily exploited for its beautiful finely streaked timber, which is used for carving, inlay work, furniture and musical instruments. The number of mature trees has declined and large parts of the habitat have been turned over to crops. Felling of the species is now allowed only by quota, but the high price of the timber drives an illegal trade.

Belian or Ironwood, a species found in lowland primary forest, is one of the most renowned timbers of Borneo. It is used locally in building and for water butts. The commercial uses of Belian are in heavy construction, boat building, industrial flooring, roofing and furniture. Belian has been valued by the Chinese for making coffins. In southern Kalimantan, Belian is felled by the owners of concession rights and also by local people co-ordinated by Ulin traders. Transmigrant settlers in East Kalimantan cut and sell these trees to supplement their income from cultivation. Unfortunately, natural regeneration of Belian is very limited once the forests have been logged and it is planted on only a small scale because the supply of seeds and seedlings is inadequate. Overharvesting has resulted in the species being listed as 'Vulnerable' by the IUCN.

Fire has also been a major threat to the forests of Indonesia. The total forest area affected by burning in 1997 is estimated to have been at least 2.4 million hectares (six million acres). An estimated 70 million South East Asian people were affected by the fire haze. The fires were caused by a combination of deliberate land clearance for commercial oil-palm and timber plantations, Government-sponsored transmigration projects and burning by local subsistence farmers. Drought conditions enabled the fires to continue into 1998, affecting areas such as the species-rich forests of East Kalimantan province. Throughout Borneo and Sumatra, Orang-utans and other primates were among the unfortunate victims.

OPPOSITE LEFT Montane rainforest can be found at altitudes of up to 3,500 metres (11,500 feet). In montane forests, the tall dipterocarps disappear and make way for smaller trees of the oak and laurel types.

ABOVE Many species of epiphytic ferns flourish within the moist rainforest environment.

89

Wildlife Under Threat

Deforestation has devastating consequences for Indonesia's wildlife. The list of endangered species is seemingly endless. In Java, a few thousand Javan Gibbons hang on precariously because more than 95 per cent of their forest habitat has been wiped out. Fewer than 4,000 Asian Elephants survive; and at most 650 Sumatran Tigers. The Javan Rhinoceros has become the world's rarest large mammal, with fewer than 60 individuals in the Ujong Kulon National Park in western Java and isolated populations in Cambodia, Laos and Vietnam. There have been intensive and at times controversial, efforts to save this species, but its future is not yet guaranteed.

The Javan Rhinoceros spends much of its time submerged in water or wallowing in mud. Dense lowland rainforest, with a plentiful supply of water and mud, is its preferred habitat. Loss of forest and illegal hunting have propelled its decline and the very low numbers mean that any threat could now be catastrophic.

The smallest and hairiest rhino, the Sumatran Rhino, clings on to survival in Sumatra, Borneo and pockets of Peninsular Malaysia, Myanmar (Burma) and Thailand. Water is important to this endangered forest species. As with other rhinos, overhunting for its horn has been a major threat and illegal poaching continues to supply the Chinese 'medicine' trade.

The 'Man of the Woods'

Perhaps the most charismatic of all Indonesia's wealth of wildlife is the 'Man of the Woods', the endangered Orang-utan. Asia's only great ape is now confined to forests in Sumatra, Kalimantan and the eastern Malaysian states of Sabah and Sarawak. Some experts consider the populations of Borneo and Sumatra to be separate species. Orang-utans are solitary animals that live mainly in trees, rarely descending to the forest floor. Fruit is their main food, together with leaves, shoots, bark, insects, nestling birds and small mammals. Orang-utans rest at night in nests built in the trees. The young live with their mothers until they are seven or eight years old.

People have exploited Orang-utans for thousands of years: for food and as a human substitute in headhunting rites. In the 1960s, capture of live animals for zoos and circuses led to further declines. Now, habitat destruction is its major enemy.

Protecting the forest habitat from disturbance and destruction is the only way to halt the decline. Wild populations suffered heavy losses during the forest fires of the 1990s, a tragic situation that helped to focus international attention on the plight of this endangered species. Every effort will be needed to maintain the Gunung Leuser National Park, the Orang-utan's last stronghold in Sumatra and the national parks that protect its forest habitat in Kalimantan.

Another well-loved primate which has become a flagship species for conservation is the somewhat comical-looking Proboscis Monkey, which is endemic to Borneo. Mangrove swamps and riverine forests are the habitat of this large leaf-eating creature. Unfortunately, areas along rivers are often the first to be cleared for agriculture and human settlement. The Proboscis Monkey can survive in areas that have been selectively logged, but fragmentation of its habitat is a growing problem.

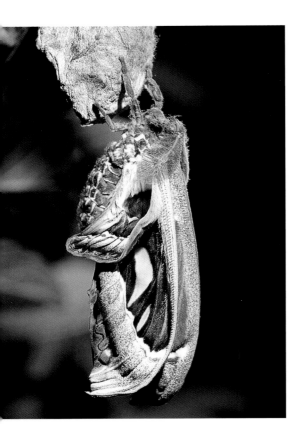

The Maleo in Sulawesi

Found only on the island of Sulawesi, the Maleo is one of Indonesia's most remarkable rainforest birds. This chicken-sized relative of the domestic fowl has a strange, helmet-like structure on its head. The Maleo is known for its curious communal nesting habits. Each female lays a single egg, which she then buries up to one metre (three feet) deep in warm sandy beaches or in forest clearings warmed by volcanically heated underground streams. Once the eggs are buried, the adults carry out no further parental care. The young hatch after 60–80 days, emerge from the ground like baby turtles and are ready to fly.

About 130 traditional nesting sites for Maleos are currently known. Most are threatened or have already been abandoned. Coastal sites have been more seriously affected by habitat destruction, as this is where development and human population growth are concentrated. Clearance for cultivation, logging and mining is also a major threat. Such habitat loss is the greatest danger to the Maleo's survival. The birds often forage several kilometres away from the nesting grounds and while the adults can tolerate having to traverse degraded forest to reach their foraging grounds, the young chicks struggle to survive where there is no intact forest connecting the nesting sites to foraging areas.

Other threats to the Maleo are hunting and egg-collecting. The large eggs are easy to find and Sulawesi people value them as food. Egg-collecting is a part of the traditional culture. Pigs, monkeys and lizards also eat the eggs. Some nesting sites are managed by the forestry department or by local communities and there is scope for them to be managed sustainably. Shooting and trapping of adult Maleos is forbidden at these sites. Other sites are harvested opportunistically by semi-nomadic indigenous people and by rattan collectors. In these areas hunting of the birds is more common.

The Maleo is the official symbol of Central Sulawesi and there are Maleo monuments and symbols all over the island. Unfortunately, few local people realize that the Maleo is unique to Sulawesi and may become extinct unless conservation efforts are increased.

Case study: The Aceh Elephant Project

The most effective conservation solutions are often those that integrate the needs of endangered wildlife with those of the local people. One such model is the Aceh Elephant Project in northern Sumatra. Aceh includes a large tract of the intact rainforest that runs through the Bukit Barisan Mountains, which form Sumatra's spine. Around 800 elephants still live here. As the forest area is relentlessly reduced by logging and fires, the elephants come more into contact with people and their crops. Acehnese people traditionally respect the elephant, but conflicts are beginning to arise.

Through the Aceh Elephant Project, detailed information is being collected on the forest habitats of the elephants and related biodiversity. The information will be used to develop better land-use planning based around elephant habitat and corridors. Meetings have been held with representatives of all the different people in the region to devise a plan for elephant conservation that takes into account the views of the Acehnese communities and those of more recent settlers. Various ways to reduce conflict between humans and

OPPOSITE LEFT AND RIGHT The Atlas Moth found in Java is one of the world's largest moths. Here it is resting on a cocoon.

RIGHT Sulawesi has 265 resident bird species, of which over one-quarter are endemic. The extraordinary Maleo is perhaps the most famous and is a celebrated flagship species on the islands.

BELOW A male Sumatran Rhinoceros grazing on leaves.

elephants are being tried in areas where elephants are ruining crops. In the longer term, alternative forms of forest-friendly land use will be developed. This will benefit not only the elephants, but also the other large mammals, such as the Sumatran Rhino and Tigers, which are at risk through continuing deforestation.

The Philippines

The familiar sequence of saddening events that begins with logging and ends with the extensive deforestation of entire countries is repeated in the Philippines, an archipelago of over 7,000 tropical islands. Luzon and Mindanao are the largest of the 700 inhabited islands. Tropical forests originally covered most of the Philippines. Today, primary lowland rainforest covers about three per cent only of its original area and the only extensive areas of forest outside national parks are on Palawan. The pressure from shifting cultivation is growing as migrant farmers move into the uplands where the remaining forests are located. Under current slash-and-burn practices, forest is cleared and farmed until the soil is completely exhausted.

The lowland evergreen rainforest that remains occurs on well-drained soils and on the lower slopes of mountains where the dry season is not pronounced. These forests are dominated by various types of dipterocarps. Lauan (*Shorea*) forest occurs in lowlands and foothills up to 400 metres (1,300 feet). At higher altitudes, Lauan-Apitong (*Dipterocarpus*) forest occurs. Another type, classified as Lauan-Yakal (*Hopea*) forest, grows on volcanic soils in small areas of Luzon, Leyte and Mindanao. Much of this type of forest has been cleared. The dipterocarp forests have been a hugely important timber resource, especially since the Second World War and forest with a high proportion of mature trees is now scarce.

Other remaining forest types in the Philippines include tropical montane and subalpine forests. Small areas of forest on limestone are dominated by the valuable hardwood species Molave. Logging has caused the disappearance of Molave forests in most of the country; of the seasonally dry, monsoon forest on the western side; and of the pine forests, with *Pinus kesiya* and *P. merkusii*, in the uplands of north and west Luzon and Mindanao.

Mangrove forests occur on Palawan and parts of Luzon, Visayas and Mindanao. Palawan has 38 per cent of the remaining mangrove of the Philippines.

OPPOSITE LEFT Dense lowland forest of Bataan National Park in the Philippines.

ABOVE A rhododendron of the Sudirman Mountains, Irian Jaya. Many new species of plants await to be described in this relatively unexplored part of Indonesia.

The timber trade

Logging in the Philippines wreaked havoc in the forests from the end of the Second World War until the 1970s. Japanese demand for timber, beginning in 1961, increased the damage. The Philippine timber boom of the 1960s produced vast profits for the timber companies and the Government was unable to implement effective logging controls. A selective ban at last succeeded in reducing the export of logs between 1976 and 1981 and destructive timber exploitation decreased during the 1980s because of lower world prices, forest depletion and conservation policies. Higher log export quotas were allowed during 1982–84 because of a critical need for foreign exchange, but by 1986 the export of timber had fallen to a mere five per cent of total export income, compared to 33 per cent in 1968. An export ban on logs and unworked timber has subsequently been introduced.

Logging has been totally banned in parts of Luzon, Catanduanes, Masbate, Leyte and Negros as well as on small islands with an area of less than 500 square kilometres (190 square miles). In 1989, a bill was passed banning logging in all

LEFT The Long-tailed or Crab-eating Macaque of South East Asia is usually found at the forest edge, in swamps or in coastal forest.

RIGHT Male Sailfin Lizards can reach 1 metre (3 feet) in length, females somewhat less. In adults, especially males, the sail-like fin is unmistakable.

but nine of the Philippines' 73 provinces. This legislation allowed logging only in provinces that are at least 40 per cent forested. Nevertheless, illegal logging and export of timber remain a serious problem.

Natural vegetation on Palawan, the third largest of the islands, is mostly lowland monsoon and semi-evergreen rainforest. With 50 per cent of its forest cover remaining, Palawan is considered a priority for conservation both within the Philippines and internationally. The forests harbour four endemic bird species, including the Palawan Peacock-pheasant, which has been heavily shot and trapped. They are also the stronghold of the threatened Philippine Hawk-eagle, the Philippine Cockatoo and the Blue-naped Parrot. The entire province of Palawan has been declared a Fauna and Flora Watershed Reserve and contains protected areas of other types.

Despite all this, rapid development and large-scale immigration have continued apace in Palawan. The remaining rainforests are being put at risk by logging, shifting cultivation and mining. There are few valuable

dipterocarps in Palawan's flora and large-scale logging has, therefore, been commercially viable only over the past 25 years or so. Nearly all of Palawan's forests are now leased to logging operations, but most concessions have not yet been extensively worked.

The Philippines have over 3,500 tree species, of which Philippine mahogany is the most important commercial timber. This consists of wood mainly from six dipterocarp species. The timber boom of the 1960s and 1970s inevitably put the heavily traded timber species under pressure. Peter Ashton, the leading authority, considers that all 40 or so dipterocarp species in the Philippines can be classified as globally threatened.

Mangkono yields the hardest of Philippine timbers. The species is naturally rare, being restricted to the islands of Homonhou, Dinagat and the eastern tip of Leyte. The hardness of the wood gives the trees some protection against felling but, following the depletion of other hardwoods, local people are cutting down the smaller trees and conservation measures for Mangkono are urgently needed.

Endangered species

The Philippine or Monkey-eating Eagle is one of the world's most endangered birds of prey. This huge rainforest bird, the second largest eagle in the world, still occurs throughout its historical range on Luzon, Leyte, Samar and Mindanao, but only a few individuals remain. In the past, hunting and trapping were the main threats, but habitat loss is now a greater danger. The Philippine Eagle feeds on snakes, flying lemurs, flying squirrels and, occasionally, monkeys.

Endemic and threatened mammals of the Philippines' forests include the Dwarf Water Buffalo or Tamaraw of Mindoro island. The Tamaraw grazes in grassland areas and uses the forest for shelter. Fewer than 200 individuals are now thought to remain in the wild. Traditionally considered to be fierce, Tamaraws have been shot virtually to extinction by hunters keen to prove their own bravery. Habitat destruction has also taken its toll, as Mindoro has lost most of its rainforest. Efforts to conserve the Tamaraw have concentrated on the Mount Iglit Bako National Park, home of the largest remaining wild population. A captive breeding programme established in the park in 1979 was not a success, however. The captive population has been moved to a more secure site on the island, but the Tamaraw's future remains uncertain.

Another endangered endemic rainforest mammal is the Philippine Tube-nosed Fruit-bat, which is restricted to the southern part of Negros Island. Breeding populations of this fruit-bat are found only on the slopes or summits of forested mountains and, unless remaining areas of forest in the area are protected, this species faces extinction.

Conservation measures

The Government of the Philippines has introduced various measures to combat the depletion of forest resources through timber exploitation, but illegal logging and export of logs continue to be a major problem. In the late 1980s, one expert claimed that 'a major constraint in the management of natural forests in the Philippines is the inability of the Government to control illegal logging anywhere, including national parks.' The other important challenge is severe rural poverty, which is an underlying cause of deforestation in many parts of the world. Protected areas have been declared under various forms of legislation in the Philippines, but there is no agreement on how many legally protected areas actually exist.

Despite the huge problems of forest conservation in the Philippines, there are some positive success stories. In 1992, the Government established a National Integrated Protected Areas System with financial support from the Global Environment Facility (GEF). This initiative concentrated on improving management and protection for 10 priority sites. One of the most important of these is the Northern Sierra Madre Natural Park on the island of Luzon, an important site for the Philippine Monkey-eating Eagle. This protected area now constitutes the largest area of conserved primary rainforest in the Philippines.

LEFT This Swallowtail Butterfly (*Papilo palinurus*) is a widespread species of South East Asian rainforests.

OPPOSITE TOP Untold numbers of beetle species inhabit the world's rainforests, each with their own important ecological role.

OPPOSITE BOTTOM The Monkey-eating Eagle is a magnificent flagship species for rainforest conservation in the Philippines.

Another success has been on Cebu Island, where, against all odds, the wildlife still has a chance of survival. At least 11 of the island's endemic birds were thought to be extinct. However, a pleasant surprise came in 1999 when a survey revealed that all but two of these could still be found. The survey also confirmed considerably more surviving forest habitat than had previously been recorded. Conservation attention has focused on the Tabunan Forest, a remnant patch of 100 hectares (250 acres) which is home to the celebrated Cebu Flowerpecker. This beautiful bird was believed to be extinct until its rediscovery in 1992. It is also now known to occur in at least two areas other than the Tabunan Forest.

Local conservationists are working in partnership with Fauna & Flora International and other international agencies. The hope is that continuing research and monitoring, along with reforestation, education and training will secure the long-term conservation of Cebu's precious forests and the threatened creatures that depend on them.

Chapter Six

Central America

A Rainforest Corridor

Toucans, exotic blooms, turtles and fish embellish the
vividly coloured *molas* worked by Kuna Indian women in Panama.
These are the images of the Central American rainforests. Intricately
interspersed with geometric patterns, these representations of
rainforest wonders are a special component of the hand-embroidered
and appliqued artworks that are turned into clothing and wall hangings.
Retaining their traditions and legends, the Kuna Indians continue
to live in harmony with their rainforest environment. They were
the first indigenous people in the world to set up an internationally
recognized rainforest protected area and are an inspiration to
rainforest protection groups around the world.

Mexico to Panama

The rainforests of Central America form a corridor from the central and southern states of Mexico through to Panama, connecting with the Choco forests of Colombia to the south. In most of the six Central American countries south of Mexico (Belize, Costa Rica, El Salvador, Guatemala, Honduras and Panama) only tiny fragments of rainforest now remain.

Belize, which remains sparsely populated, is the exception. The forest remnants of Central America are extremely rich in species and are of great ecological and economic importance in this turbulent region. Yet these irreplaceable forests are being cleared at a rate that is truly alarming.

Central America is a mountainous region where montane rainforest and cloud forest clothe the volcanic slopes. Lowland rainforest is found along the Caribbean coast, where the most extensive remaining areas are in Belize. The climate is more arid along the Pacific coast, where dry deciduous forest is the natural vegetation.

Destruction of the forests

The rate of deforestation in Mexico and Central America has been dramatic, particularly over the past 50 years. The general pattern of forest loss over this period has been the opening up of forests by road building, followed by the arrival of settlers and the development of ranching. Roads have been built in the rainforests to speed up national development and to provide access for the military, for oil exploration and for logging. The building of the Pan American Highway has opened up forest areas in the region. Originally conceived in 1923 as a single inter-continental route connecting North and South America, this major project has entailed building more than 25,000 kilometres (16,000 miles) of continuous road from Alaska to the tip of South America. The only missing link is a stretch of about 80 kilometres (50 miles) through the Darien Gap, a dense area of rainforest in Panama and neighbouring Colombia. The threat of road development and settlement farther north in Panama led the Kuna Indians to take control of their forestlands.

Road building is usually followed by the arrival of settlers, who clear the forest to grow food for their own use, such as maize, manioc, beans and rice, together with crops for sale. Much of the forest soil is shallow and of low fertility, unsuited

PREVIOUS PAGE The Monteverde Reserve protects an important area of Costa Rica's cloud forest.

LEFT Sugarcane cultivation in northeastern Belize is now one of the main reasons for deforestation in this part of the country. Much more extensive areas of Belizean rainforest were cleared for agriculture by the Mayans 1,000 years ago, but subsequently recovered with the decline of the Mayan civilization.

RIGHT This climbing vine saves energy by using the tree for support – a common ecological strategy in the rainforest.

to intensive agriculture. After two or three years, the settlers move on to clear new areas. The degraded land cannot recover because it is sold to speculators and for cattle ranching. Clearance of forest for cattle ranching – the so-called 'hamburger connection' – was a campaigning issue for environmentalists in the 1980s.

The conversion of forest for plantation agriculture has been another major threat. There are banana, coffee and oil-palm plantations throughout the region. Fortunately, coffee is now beginning to be grown in forest-friendly schemes, such as the Rainforest Rescue scheme described in the case study later in this chapter. This is unusual, however.

A Wealth of Plants

Many new plant species await discovery in Central America. Although still little known botanically, the region is certainly among the richest areas of the world for plant diversity. The forests yield many economically important plants, including precious timbers, orchids and the wild relatives of some of our major crops. The original source of chewing gum, the tree *Manilkara chicle* is found within the region. In the forests of southern Mexico, Guatemala and Belize, about 5,000 *chicleros* still collect *chicle* using traditional harvesting methods.

A salad favourite, the avocado, *Persea americana*, originated in Central America and it has been cultivated in the area for several thousand years. Wild avocados and other closely

related species still grow in some parts of the rainforest. The relatives of the wild avocado are considered to be important genetic resources for the future. Three such species are found in an area set aside for nature conservation near Purulha, Baja Verapáz, in Guatemala, a reserve which has been set up to conserve the endangered Resplendent Quetzal, one of the world's most beautiful birds. Wild avocado plants are an important source of food for the Resplendent Quetzal.

The Quetzal is the national bird of Guatemala. According to legend, the male's crimson coloration has been present only since the time of the Spanish colonization. Previously all-green, the male birds are said to have swooped down to protect the Mayan Indians in battle and were stained with blood in the process.

Maize was domesticated in prehistoric times in Mexico and Central America and is of cultural importance throughout the region. A wild species (*Zea perennis*) was thought to be extinct in the wild but was rediscovered in 1977. A new species (*Z. diploperennis*) was also found and is now protected in the Sierra de Manantlan Biosphere Reserve in Mexico.

Important timbers of Central America include two species of mahogany. *Swietenia humilis* is distributed along the west coast from southwestern Mexico to Costa Rica. This dry-forest species is no longer of commercial value, whereas *S. macrophylla*, which occurs from southern Mexico, through Central America to Bolivia and Brazil, is one of the world's most valuable commercial timbers. It is now considered to be vulnerable, mainly because of overexploitation. Logging for international markets has exhausted supplies, particularly in the northern parts of its range. The ecology of the species makes mahogany vulnerable to logging regimes. Harvesting and processing are generally only 50 per cent efficient. There is, at present, little economic incentive to manage natural stands sustainably. International trade in timber of the species is subject to the provisions of Appendix III of CITES.

Honduras Cedar is another important timber tree that is now vulnerable after 200 years of large-scale exploitation. Large individuals have become scarce and the species is now widely threatened. Trees are often cut opportunistically in the search for other species such as mahogany.

Sustainable forest management for these and other species is developing in the region, but much more needs to be done.

LEFT A rich variety of plants is concentrated in the dwindling rainforests of Central America. Many of the species provide medicine, food and fibres, as well as wood.

TOP RIGHT A beautiful but rare bird of the Central American forests, the Resplendent Quetzal was worshipped by both the Mayan and the Aztec civilizations.

RIGHT Costa Rican Tarantula spiders tend to burrow and will feed on crickets and other large insects, as well as the occasional baby mouse. At the slightest threat of danger, these spiders can display incredible speed.

Endangered Animals

The largest tropical cat of the Americas is the Jaguar. Once quite common from California to Argentina, this solitary and handsome animal is now extinct in North America and also in most of Central America.

Jaguars are good at swimming, running and climbing trees. They are opportunistic feeders, catching aquatic animals such as turtles and capybaras and also taking monkeys, peccaries and birds. Hunting for its prized spotted fur has, in the past, been the major threat to the Jaguar. Legal protection through CITES has virtually eliminated the commercial trade in furs and Jaguars are now more often shot by farmers because they prey on domestic animals. Belize, which has the largest remaining population of Jaguars in Central America, has created the Cockscomb Basin Jaguar Preserve in the Maya Mountains specifically to protect the species.

The national animal of Belize, Baird's Tapir is a shy, generally nocturnal creature that lives near water, in mangrove, lowland and montane forests. It has a wide geographical range, from the forests of southern Mexico to Ecuador, but is fast declining through habitat loss and hunting for meat. The best hope for the survival of this vulnerable mammal is the protection of the remaining extensive forest blocks in Central America. One of the largest of these is in northeastern Honduras and northern Nicaragua: an area

where human colonization and forest clearance continues. A 1994 survey concluded that there could be up to 2,700 tapirs in the area. The Government of Honduras has proposed the establishment of seven conservation areas covering over 10,000 square kilometres (3,900 square miles) of forest, but, unfortunately, farmers and ranchers have ignored the boundaries, or are unaware of them, and land clearance continues apace.

The Central American forests shelter spectacular birds such as the Resplendent Quetzal, the Scarlet Macaw, the Military Macaw and the critically endangered Horned Guan, a large turkey-like bird with a remarkable red, knobbly horn on the top of its crown. Once quite common in the montane rainforests of southern Mexico and Guatemala, Horned Guans are now thought to number less than 1,000 because of hunting and habitat loss.

Conservation in progress

Despite the huge problems and pressures in Central America, conservation efforts are making progress. Every country now has protected areas which, in Costa Rica and Guatemala, amount (at least on paper) to over 20 per cent of the land area. The largest protected area in the region is the Maya Biosphere Reserve in northern Guatemala. Half the area of

LEFT Capuchin monkeys are found in the tropical forests of Central and South America. Generally, Capuchins are active during the day, mainly in the forest canopy, where they feed on fruit and small animals.

OPPOSITE TOP Jaguars – the dappled coats of these magnificent cats blend with the light and shade of the forest floor.

OPPOSITE BOTTOM Blue-crowned Motmot. The nine species of motmot are confined to continental South America. A motmot will wait motionless for its prey – insects, small frogs, lizards and snakes – and then dart swiftly to catch it.

this reserve is a multiple-use zone, where sustainable harvesting of *chicle*, palms and other useful plants takes place. This allows local people to maintain their traditional lifestyle. Five large core zones of the Maya Biosphere Reserve are strictly protected.

La Amistad National Park is a trans-border protected area in the Talamanca Mountains of Costa Rica and Panama. Here, the largest undisturbed stretch of cloud forest in Central America harbours Jaguars and Harpy Eagles. Of global importance, La Amistad is also a birdwatcher's paradise, visited by 75 per cent of all the migratory birds of the Western Hemisphere.

Another exciting regional initiative is the development of the Mesoamerican Biological Corridor (MBC). The aim of MBC is to create an ecological route by protecting a series of priority areas. Existing protected areas are being strengthened, new protected areas created and connecting areas managed in a biodiversity-friendly way, through reforestation and agroforestry schemes. The Maya Biosphere Reserve forms part of the MBC, connecting directly with the Calakmul Biosphere Reserve in Mexico and the Rio Bravo Protected Area in Belize.

Case study: Rainforest Rescue

The Rainforest Rescue Programme of the National Arbor Day Foundation is helping to conserve the biologically diverse agroforestry systems of the traditional coffee farms in and around the El Triunfo Biosphere Reserve in the state of Chiapas, Mexico. El Triunfo is a 119,000-hectare (294,000-acre) reserve, protecting various forest types including lowland rainforest and cloud forest. The aims of the project are to integrate ecosystem conservation into sustainable agricultural activities, to diversify coffee fields by planting useful native trees and to improve marketing of eco-friendly coffee. This project forms part of Conservation International's larger Sustainable Coffee Initiative. Also in Chiapas, Rainforest Rescue is working with the Nature Conservancy and a local non-governmental group, the Instituto de Historia Natural de Chiapas (IHN), to secure a community-based conservation programme at the El Ocote Forest Reserve. As part of this programme, training and technical assistance is being provided for the production of organic coffee by a group of 12 *ejido* (communal) farm groups.

Eco-tourism

Eco-tourism is increasingly important throughout Central America. Costa Rica has recognized the economic benefits of tourism for many years and is well known as an eco-tourism destination. Traditionally, Costa Rica attracted tourists from other Central American countries, especially Nicaragua, but over the past two decades visitors have come increasingly from North America and Europe. About one-third of Costa Rica's travel agencies now specialize in eco-tourism. A popular rainforest destination is the Corcovado National Park. Located in one of the wettest regions of Costa Rica, Corcovado has around 500 different tree species and an amazing diversity of wildlife including 300 species of birds, 139 mammals and 116 species of amphibians and reptiles. Corcovado protects Costa Rica's largest population of the splendidly coloured Scarlet Macaw, as well as the endangered Jaguar and Baird's Tapir.

Another internationally renowned reserve in Costa Rica is the Monteverde Cloud Forest. This non-governmental reserve of 10,000 hectares (25,000 acres) in the Tilaran Mountains of northern Costa Rica was established in 1972. Fifty years ago, Monteverde's forests were virtually untouched. Then, increases in the local population and changes in the law governing land use resulted in the expansion of agriculture and forest clearance. In the late 1950s, a Quaker community from North America came to settle in the area. They set aside some of the forest land that they had purchased for watershed conservation and developed dairy farms. The rich cloud forest at high altitudes remained largely intact and began to attract an increasing number of biologists and students. Studies led to the discovery of many rare species, including the endemic Golden Toad. As the importance of Monteverde became clear,

LEFT One of many amphibians found in Central America, this male Harlequin Frog is in fact a toad. His multi-coloured skin warns predators of his poisonous skin secretions.

ABOVE The lush rainforests of Costa Rica attract eco-tourists from around the globe.

RIGHT Many rare butterflies can be found in forests from Guatemala to Panama.

the Quaker settlers joined forces with local conservationists and scientists from the tropical science Centre of San Jose to form the Monteverde Cloud Forest Reserve.

Visitors to the Monteverde Cloud Forest are attracted by the lush forest, with palms, ferns, bromeliads and immense oak trees together with the great diversity of animal species. As well as the Golden Toad, known only from Monteverde, the reserve is famous for a variety of other rare species including bellbirds, umbrellabirds, macaws, quetzals, tinamous and agoutis.

Chapter Seven

The Caribbean
Fragile Fragments

Christopher Columbus, who first set foot in the Caribbean more than 500 years ago, would be unlikely to recognize much of the landscape to be seen there today. Most of the natural forests that he described with such a sense of wonder are no more. They have been exploited for timber, cleared to make way for plantations and subsistence agriculture, or chopped down to make space for housing and industrial development. Yet among all this destruction, precious remnants of rainforest still cling on, serving as a reminder of what Columbus saw. They still provide a habitat for a vivid array of colourful plant and animal species that the early explorers may have known. These fragments that remain are of immense biodiversity importance and give character to the islands as a whole.

Island Forests

The most luxuriant forests of the Caribbean islands are at low elevations, up to about 300 metres (1,000 feet). The dense lowland rainforests found in mainland Central and South America do not exist on the islands, although many of the majestic timber species of the mainland are found in the Caribbean forests. Natural lowland forest is generally found today only in areas that are inaccessible or unsuitable for cultivation. At higher altitudes, various types of montane rainforest still occur.

As an introduction to the variety of rainforests on a relatively small scale, the Caribbean islands have much to offer and eco-tourism facilities are increasing. Tourism is already the main source of foreign exchange for most of the islands, but relatively few tourists venture away from the beaches and water-sports areas.

Valuable hardwoods were one of the first commodities that attracted traders to the Caribbean. Early Spanish explorers chose mahogany to repair their ships and it has been traded around the world since the 16th century. It was taken as a gift to King Philip II of Spain in 1584. Lignum vitae was another valuable timber. This wood, one of the world's heaviest, has had a whole host of uses and its medicinal resin is still available in world trade.

Sugarcane was the first plantation crop on the Caribbean islands. It was taken to the island of Hispaniola on Columbus's second voyage in 1493 and sugar was exported from the island to Spain about 15 years later. Subsequent crop introductions that have led to the loss of rainforests include cotton and tobacco, established in Barbados by 1640.

PREVIOUS PAGE The herb layer of the forest floor consists of a diversity of shade-loving plants.

LEFT Sugarcane has been cultivated in the Caribbean since the earliest days of European settlement. Clearance for plantation crops has been one of the main reasons for deforestation.

RIGHT Trafalgar Falls, Dominica – one of the attractive natural features of this rainforest island.

Montserrat and Dominica

When Christopher Columbus named the tiny island of Montserrat in 1493, it was covered in lush tropical rainforest which provided shelter and food for the native Arawak Indians. For the next 130 years or so, the Arawaks continued their traditional way of life, but then, in a period of rapid change, they were displaced first by English and Irish settlers and then by the arrival of African slaves.

Within 50 years, two-thirds of the island's rainforests had been cleared to make way for sugarcane plantations. Now, much of the island is forested, but with a very different tree cover from the original forests, including many introduced species. Montserrat suffered a period of natural turbulence during the closing years of the 20th century, when dramatic volcanic eruptions and hurricanes transformed the landscape of the island.

Despite its small size (102 square kilometres/39 square miles), Montserrat has a diverse fauna and flora that includes two endemic plants, one endemic bird (the Montserrat Oriole) and three endemic lizards. It is also home to one of only two remaining populations of the Mountain Chicken, which is in fact one of the world's largest frogs.

About the size of a Guinea Pig, the Mountain Chicken is regarded as a local delicacy, the flesh of its long hind legs reputedly tasting like chicken. At one time it was found in the forests of at least five Caribbean islands, but it survives now only on Montserrat and Dominica. Hunting, habitat loss and competition with invasive species have all contributed to the extinction of the Mountain Chicken throughout much of its former range.

Another island in the Lesser Antilles, the Commonwealth of Dominica, is capitalizing on its natural heritage to attract eco-tourists from around the world. Promoted as the Nature Island of the Caribbean, Dominica has retained a far greater percentage of pristine natural habitat than many of the other Caribbean islands. Still with 60 per cent forest cover, Dominica has two national parks (Morne Trois Pitons and Cabrits) and two forest reserves.

Rich tropical vegetation is among the main attractions of the Morne Trois Pitons National Park, whose cloud forest and lower-altitude rainforest, lakes and waterfalls provide habitat for abundant bird life. One of the most visited sites in this park is the Emerald Pool, a popular spot for bathing in the rainforest.

Another of Dominica's main attractions is the Carib Indian reservation, in the northeastern part of the island. The Caribs were one of the original Indian groups that inhabited the Caribbean islands before European settlement. Fewer than

2,000 descendants of these indigenous people live within the Dominican reserve today.

Dominicans value the revenue from tourism, but many also consider it very important to preserve their cultural heritage. Great efforts are being made to ensure that local customs and traditions are respected and celebrated in the face of the expanding tourist industry.

Jamaica

Beaches, music and a carnival atmosphere are the usual images which come to mind when thinking of the larger Caribbean islands such as Jamaica. However, away from the developed areas and beach resorts there is an abundance of wildlife. The forests of Jamaica, for example, are exceptionally rich in tree species.

About 25 per cent of the original forest cover remains and two areas have been singled out for international conservation recognition. The Blue and John Crow Mountains in the east of the island support relatively undisturbed montane rainforest with a diversity of more than 600 plant species. Over 80 of these plants grow nowhere else in the world. Lush mosses, orchids and bromeliads thrive in the sheltered, humid atmosphere among the tangle of trees. Two bird species that occur throughout the area are the Red-billed Streamertail and

the Black-billed Streamertail. Both these birds feed on nectar and small insects.

Conservation of the forests of the Blue and John Crow Mountains is extremely important, not only to safeguard their wildlife, but also to protect the watershed of the agricultural lands of the foothills and the water supply to Jamaica's capital, Kingston. Deforestation for cultivation remains a significant threat, particularly on the southern slopes of the Blue Mountains, but fortunately the Blue and John Crow Mountains are now protected as a national park.

Another exceptional area of species-rich forest is the sparsely populated Cockpit Country, farther to the west. Here, rugged limestone terrain and a lack of surface water have helped prevent forest clearance for timber exploitation, agriculture or human habitation.

LEFT The Red-billed Streamertail, Jamaica's national bird, is famed for the male's extraordinarily long tail feathers. The species occurs only on the island, where fortunately it is quite common. Jamaica has the highest number of endemic bird species of any Caribbean island.

ABOVE The Scarlet Ibis is found in the northeastern coastal areas of South America and in Trinidad. Its brilliant red plumage is derived in part from the crustaceans that form an important part of its diet.

RIGHT The *Anolis* lizard is a tree lizard which forages for insects at the slender ends of branches. It is well camouflaged on tree trunks.

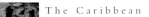

Cuba

Biologically richest of all the Caribbean islands is the largest, Cuba. Less than 20 per cent of the land here is now forested, but the fragments remaining are of global significance because of their species diversity.

Typical lush Cuban rainforest, rich in tree ferns and epiphytes, occurs at altitudes below 400 metres (1,300 feet) along river valleys in the northeast of the island. Seasonal lowland rainforest is more widespread, but little remains undisturbed and much has been converted to agriculture. Expanding cocoa, coffee and tobacco production are serious threats to the rainforest while in drier forest areas logging, shifting cultivation and charcoal production are the major worries.

Various types of montane forest occur in Cuba, with cloud forest at high altitudes in the Sierra Maestra, Pico Turquino and Pico Bayamesa mountains in the southeast. These cloud forests fall within a current large-scale conservation and sustainable-development project, the Gran Parque Nacional Sierra Maestra. Mangroves now account for around one-quarter of all Cuba's forests.

Cuba has about 6,200 native plant species, over half of which are endemic to the island. Nearly all the amphibians are endemic (36 out of 41 species) and 79 reptiles are similarly found nowhere else in the world.

Some areas of dense montane rainforest are home to an endangered mammal called the Cuban Solenodon, which has a very long flexible snout that makes it look rather like an outsize shrew. Solenodons spend most of their time foraging in the undergrowth, using their claws to dig into bark and

OPPOSITE LEFT Cuba still retains some 20 per cent of its rainforest cover and is becoming a popular eco-tourism destination.

LEFT Cuban Tree Frog. Males reach a maximum size of 5-6 centimetres (2.5 inches) while females far exceed them with their 12.5 centimetres (5 inches) length. They will voraciously eat anything that will fit in their mouth, from large cockroaches to mice.

BELOW The St Lucia Parrot is an important flagship species for the island. Rainforest is the habitat of this charismatic bird.

forest debris on the forest floor, whilst their snout helps them to locate prey. The Solenodon's decline is mainly due to deforestation. The species is legally protected and its habitat is included within several of Cuba's protected areas, thus its future seems more certain.

Cuba has 159 breeding bird species, over 20 of which are endemic. The Giant Kingbird is one of the endemic species that is endangered. The Ivory-billed Woodpecker, once widespread in virgin forest of the USA as well as Cuba, is now believed to be extinct. The last possible evidence of its existence – damaged tree bark – was found in Cuba in 1991.

Birds found elsewhere in the Caribbean forests include the Amazon parrots. The islands of St Vincent, St Lucia and Dominica all have endemic parrots that have become flagship species for conservation. These three species are all strikingly attractive, well known to local people and dependent on undisturbed forest habitats for their survival. The St Lucia Parrot, despite legal protection for over a century, was close to extinction 20 years ago, with only 150 individuals remaining. The main reasons for the decline of these colourful and exotic creatures are habitat loss, hunting for local use and capture for the international trade in live birds. Fortunately, captive breeding and a re-establishment scheme have helped to reverse the trend.

Chapter Eight

The Amazon

The Place and its People

Words like vast and awe-inspiring barely hint at
the mystery and wonder of the Amazonian rainforest,
a region of – above all – exuberance of life. Charles Darwin,
on his South American travels, wrote of the beauty of the flowers,
the glossy green of the foliage and the unbelievable luxuriance
of the vegetation. He was filled with profound admiration.
More than a century later, only a tiny fraction of the Amazon's
unique and rich biodiversity has been catalogued. Vast
areas remain largely unknown. Images of Amazonia continue
to inspire, but just as the world is at last beginning to wake
up to the global importance of the Amazon rainforest,
this huge and irreplaceable treasure is being
relentlessly felled and despoiled.

Forest Ecology

The Amazon Basin is a region of superlatives. More than half of the world's remaining tropical rainforest is here. More species of plants and animals live here than anywhere else on Earth.

The Amazon's biodiversity is almost too great to comprehend. About one-third of the world's total number of flowering plants grow here: an estimated 80,000 species. Insect diversity would almost certainly be just as impressive – if it were possible to measure it with any accuracy. So many species have yet to be described. The Amazon, the world's second longest river, holds about one-fifth of all the fresh water on the planet. This gigantic river system contains an estimated 2,000 freshwater fish species, compared with about 300 in the whole of Europe. Such richness makes it difficult to accept the grim truth that the Amazon Basin is experiencing a higher rate of forest destruction than anywhere else in the world.

The Amazon has its sources in the Peruvian Andes, less than 90 kilometres (56 miles) from the Pacific coast. The river falls almost 5,000 metres (16,400 feet) in the first 1,000 kilometres (620 miles) of its course, but thereafter the vast river system is almost completely flat. Two-thirds of the Amazon Basin is in Brazil, but it extends into Argentina, Bolivia, Colombia, Ecuador, French Guiana, Peru, Surinam and Venezuela.

The extensive floodplains of the Amazon and its tributaries cover about 100,000 square kilometres (38,600 square miles) – almost two per cent of the Amazon Basin. The floodplains support *varzea* and *igapó* forests, each with its own unique ecology. *Igapó* forests are flooded by clear, nutrient-poor water, while *varzea* forests are flooded by rivers carrying relatively nutrient-rich water. A large number of fish species feed on the seeds of varzea trees, thereby helping to distribute the seeds. During floods, millions of fish are carried into the forests, where they feed on fallen leaves and fruit, returning to the rivers when the floods subside. Some species have specially adapted teeth, similar to those of grazing mammals, and strong jaws capable of cracking Brazil nuts. As many as 200 fish and tree species are interdependent in this way.

PREVIOUS PAGE Upland or *Terra Firma* Amazon Rainforest, Serra Dos Carajas, South Eastern Brazil.

LEFT The fruit of the Urucu contains seeds that are used as body paint. It is believed to give magical protection from the spirits of the forest.

ABOVE Spectacled Caimans are nocturnal predators and grow up to 3 metres (10 feet) long. They are fierce carnivores feeding on a variety of animals, from insects to mammals.

RIGHT Piranha, with their fearsome reputation, are typical species of the Amazonian *igapó* forest.

The unflooded areas of the Amazon forest are known as *terra firma*. These rainforests are extremely species–rich and diverse. They are much more extensive than the flooded forest areas. The terra firma forests include both dense forests, found in areas that do not have extremes of waterlogging or drought and more open forests, in areas where the climate is more seasonal or the drainage poor. Some of these more open forests are rich in palms such as the Babacu, which is very valuable and also the Brazil nut tree. Amazonian montane forests are found on the Guayana highlands.

Man and the Rainforests

Often thought of as pristine jungle, the Amazon has in fact been inhabited by human beings for about 10,000 years. Archaeologists have discovered evidence of complex civilizations along both the Amazon and the Orinoco.

When European colonists arrived in the 16th century, they found Indian settlements all along the upper reaches of the Amazon. Francisco de Orellana was impressed by the agricultural activities that he saw during his epic journey down the Amazon in 1542. However, changes to the Amazon forests were gradual until the onslaught of the past 40 years. Until the 20th century, the small number of settlers who moved deep into the forest made their living mainly by collecting and trading natural products, such as Brazil nuts, cocoa, cinnamon, precious woods, river turtles and fish. The descendants of these early settlers are the *caboclos*, usually of mixed Indian and Portuguese descent. Attempts to clear rainforest for commercial agriculture during the 19th century were largely unsuccessful. Shortage of labour was one of the reasons, especially after the abolition of slave labour in 1888.

Rubber: the beginnings of prosperity

The Rubber Tree *Hevea brasiliensis* is native to the Amazon forests. It was long known to the native Indians, originally as a source of food. After treatment to remove poisons, its seeds were eaten and they are still used as fish bait in parts of the Amazon today. The Omagua Indians, a once-powerful tribe living in the Upper Amazon area, used rubber to make water bottles, balls and elastic bands for the use of the tribe and for trade.

Exploitation of rubber for international markets began in earnest in the mid 19th century, spreading inland from the mouth of the Amazon to the forests of Bolivia, Colombia and Peru. The cities of Manaus and Iquitos flourished as a result of prosperity from rubber. Latex extraction and processing was carried out mainly by seringueiros from northeast Brazil. The rubber barons made great fortunes, especially when the bicycle craze of the 1890s hugely increased demand for the material for tyres. Meanwhile, the migrant seringueiros lived a hand-to-mouth existence.

LEFT The Kinkajou, sometimes known as the Honey Bear, lives in the forests of Central and South America. It is almost entirely arboreal, spending much of the day in a hollow tree and emerging at night to feed on fruit, honey, insects and other small animals.

ABOVE Amazon boys enjoying life in the rainforest.

Brazil tried to retain a monopoly on rubber, but eventually, by deception, Rubber Tree seeds were taken from the country. Henry Wickham was given permission to take seeds to Kew Gardens, on the pretext that they were a gift for Queen Victoria. The seeds arrived in 1876 and germinated successfully. From Kew, rubber seedlings were exported to Sri Lanka and Singapore and formed the basis for industrial production in Malaysia. Most of the world's rubber still came from the Amazon forests until 1910, when the main source of supply switched to plantation-grown rubber from Malaysia. The decline of Amazonian rubber production resulted partly from a fungal disease, South American leaf blight. Today, South East Asian plantations account for most of the world's rubber.

Investment and settlement

Between the early 1960s and the end of the 20th century, the population in the Brazilian part of this rainforest region soared from about two million to 20 million. Much of the increase has been a result of deliberate government policy to open up the interior. The Brazilian Government launched Operation Amazonia in 1965. Tax incentives encouraged private investment, mainly in the creation of huge cattle ranches. A major wave of settlement in the Brazilian Amazon began with the building of the TransAmazonica Highway in the 1970s. The road was seen as a way to provide land for poverty-stricken people in Brazil's coastal cities. Settlers could claim productive land once they had cleared the forest vegetation. Unfortunately, they were not told how to cultivate the generally infertile forest soils and forest plots were soon abandoned.

In the early 1980s, money provided by the World Bank helped to open up the states of Mato Grosso and Rondonia with road-building programmes. Widespread colonization of Indian lands followed. It has been estimated that every kilometre of new road resulted in the destruction of between 400 and 2,000 hectares (1,000–5,000 acres) of forest. As a result of road building during the period 1972-85, deforestation in the Brazilian state of Para increased from less than one per cent to over 17 per cent.

As in Central America, abandoned unproductive farmland was generally sold on for cattle ranching, which was heavily subsidized by the Brazilian Government. In fact, the principal source of income from ranching in Brazil has often been land speculation rather than the sale of beef. Ownership of large ranches has made great fortunes for a few people in Brazil, but at an incalculable cost to the environment.

The countries of Amazonia have no tradition of forest management for commercial timber production such as that developed in parts of South East Asia and Africa a century ago. Of Amazonia's many thousands of tree species, only about 50 are widely exploited. Very little is known about the extent and abundance of most Amazonian timber species. Broad-leaved Mahogany has dominated exports and much of the timber is illegally harvested from Indian lands in Brazil. The USA and the United Kingdom have been the main importers of mahogany. In Bolivia, the greatest volume of trade has also been in Broad-leaved Mahogany, which accounts for over half the sawn wood produced in the country, together with Roble and Hura. The more accessible forests of the Amazon have generally been the most extensively logged. An unfortunate recent development has been the arrival, especially from South East Asia, of transnational corporations who are interested in large-scale logging but have no commitment to long-term sustainability of the forests. Asian logging companies now own or control over 12 million hectares (30 million acres) of Amazonian forest.

Burning has been another major threat to the Amazon forests. Ranchers and subsistence farmers set fire to the

OPPOSITE LEFT The Cock of the Rock, a bird of South American forests, has a spectacular courtship display.

OPPOSITE RIGHT A giant millipede of the Brazilian rainforest.

LEFT There are six species of spider monkey found within the tropical forests of Central and South America. They are thought to be very intelligent and use specific signals and vocal sounds to alert group members to threats from predators such as eagles, Jaguars and human beings.

BELOW The Groove-billed Toucanet occurs in the forests of Colombia and Venezuela.

vegetation on their land to create new plots for agriculture and to clear existing pasture of invasive weeds. Fire-related deforestation has increased in Brazil in recent years, coinciding with a period of drought and very high winds attributed to El Niño. In 1997 alone, for example, 1.2 million hectares (three million acres) of forest were burnt and an even larger area was destroyed in the following year. Most of the burning was on the rainforest fringes in northern Brazil. Researchers fear that the Amazon is losing its capacity to protect itself from fire.

Hunting is also on the increase throughout the Amazon. As elsewhere in the rainforests, logging roads improve access to markets and shotguns are becoming more widely available. Primates, Jaguars and Pumas, large birds, deer, tapirs, peccaries and large rodents are commonly hunted. Among the primates, Woolly monkeys and Spider monkeys are the most frequent victims.

The struggle to save the Amazon and its wildlife continues against incalculable odds.

Endangered Species

The uakaris are the only short-tailed primates found in the Americas. The closely related Red Uakari and White Uakari live in troops of 50 or more on the floodplains of the Amazon. The main reason for their decline in the wild appears to have been over-hunting, mainly for food. Both species are protected by law in Brazil and Peru.

Scientists recognize four different types of bearded saki in the Amazon rainforest. Two of them live in the southeast of Amazonia, in the Brazilian states of Pará and Maranho and are threatened with extinction in the wild because of habitat modification, deforestation and hunting. The areas where they live are now densely populated and the best hope for the survival of these primates is protection in the forests of the *latifundiàrios*, the major landowners of the region.

The Small-eared Dog or Zorro lives in lowland forest of the Amazon, Orinoco and Parana river basins. Very little is known about this species, which is protected by law in Brazil and Peru.

The Giant Otter is the largest and probably the rarest of all the world's otters. Growing to a length of nearly two metres (over six feet), this legendary creature is known to local people as the River Wolf. It is found in rivers and creeks of the Amazon Basin, preferring fairly shallow water. Giant Otters build dens along the banks of rivers and live in family groups of up to nine individuals. Unlike most other otters, the groups often fish together, with one staying behind to look after the cubs. When at rest, members of the group groom each other.

Giant Otters have a wide distribution and were once found quite commonly throughout the Amazon. Hunting has, however, been a major threat, partly because the skins were prized by the international fur trade. During the years 1946-1971, Peru supplied over 24,000 Giant Otter pelts, most of which were exported to Italy, Germany and the United Kingdom. The international trade collapsed when the species was listed in Appendix I of CITES, but poaching still continues. Giant Otters can be an easy target because they are naturally inquisitive animals. Habitat loss is another major threat, as once-remote areas of the Amazon are opened up for development. River pollution also takes its toll, the use of mercury in gold extraction being one particular threat.

The Pink River Dolphin, otherwise known as the *boto* or Amazon River Dolphin, is one of the most mysterious creatures on earth. No one knows how many botos remain in the wild, but they are still thought to be abundant. Fortunately, they are fully protected by law in Bolivia and Brazil. During the flood season, these dolphins leave the main river channels and move into the inundated forests. The second dolphin found in the Amazon region is known as the *tucuxi*. Unlike the *boto*, this species, one of the smallest of all cetaceans, inhabits both freshwater and coastal habitats.

OPPOSITE LEFT The Giant Otter – largest of all otters – lives in the rivers and creeks of the Amazon Basin. Habitat protection is vital to ensure the future of this species.

LEFT The Harpy Eagle has a huge range over the lowland rainforests, which extends from southern Mexico right down to northern Argentina, but is nowhere common.

BELOW The White Uakari Monkey has a distinctive scarlet face.

The Amazon Manatee is found throughout the Amazon Basin and the coastal waters of Brazil. It probably also occurs in the Orinoco. From the late 18th century until about 40 years ago, up to 10,000 of these large aquatic mammals were caught each year and killed to supply oil, meat and skins. In 1973, the Amazon Manatee was given legal protection in Brazil. The species is now protected throughout its range and is listed in Appendix I of CITES, but unfortunately hunting continues and the Amazon Manatee is close to extinction in Peru and other countries on the edge of its range.

The Harpy Eagle favours lowland rainforest and each pair requires a large tract of forest habitat. The eagles prey on monkeys, sloths, reptiles and large birds, which are plucked from the canopy trees. Forest destruction is the main risk for this species and the birds are illegally hunted throughout their range.

The Golden Parakeet, one of the Amazon's threatened bird species, is one of the most valuable birds in the world and is consequently a target for illegal trapping. Export of wild birds from Brazil, the only country where it occurs, is banned, but smuggling still goes on. The Golden Parakeet is also hunted for food and sport and habitat destruction remains the primary threat. Fortunately, the Tapajós National Park in Pará provides a sanctuary for this species.

Conservation Measures

As in all countries, the conservation of biodiversity competes with many other conflicting interests in the political agenda. Recent environmental legislation in Brazil has proved largely ineffective in practice. The most important international initiative is the Pilot Program to Conserve the Brazilian Rainforest. This is designed to channel funds from major industrial countries into research, land-use planning and the development of reserves and other conservation programmes.

Extractive reserves

In parts of the Amazon, native Indians and *caboclos* are working together to conserve the forests on which they depend. *Caboclos* live by farming, fishing, hunting and collecting rubber and Brazil nuts for sale. The *caboclos* have always lived hard lives, controlled by the local patrons. Many have been forced to move to the cities or find work in the mines. Those who remain are demanding a better life.

Extractive reserves are being developed as a new way to manage the forests fairly and sustainably. In these specially designated areas, the forest dwellers are allowed by law to collect nuts, rubber and other natural products. They can decide collectively how to manage their land. The change has not been easy, however and has been marked by violence. Chico Mendes, a leader of the rubber-tappers, was shot and killed in December 1988. The first extractive reserve was created the following year. It covers 500,000 hectares (over one million acres) of rainforest in the Brazilian state of Acre. Other extractive reserves have since been established in the Brazilian Amazon.

Protected areas

In addition to the extractive reserves, large areas of the Amazon rainforest have already been designated national parks or other forms of protected area for long-term conservation. In total, roughly 10 per cent of the forests throughout South America are protected. Venezuela is the country with the highest proportion of protected forest: around 25 per cent. The figure for Brazil and Peru is less than 10 per cent.

Some of the national parks that have been declared provide excellent opportunities for eco-tourists. The Manu National Park is one of the top rainforest destinations for visitors to South America. It covers 1.8 million hectares (4.4 million acres) of Peruvian rainforest, making it one of the largest protected areas in the world. It is a stronghold for the Giant Otter and other animals and birds such as Giant Anteaters, tapirs, squirrel monkeys, macaws and toucans.

OPPOSITE LEFT Igapó flooded rainforest along the Negro river, Amazonas State, Brazil.

ABOVE Varzea flooded rainforest, during the flood season, Amazonas State, Brazil.

LEFT Henry Bates, the Victorian naturalist, collected over 700 butterfly species near his home in Pará. The total number of butterfly species in the Amazon is unknown.

People of the Forests

About 140 different tribal groups still live deep in the Brazilian rainforests. In Brazil, as a whole, Indian territories cover nearly 800,000 square kilometres (309,000 square miles). In Peru, 1,000 Indian groups occupy around 740,000 square kilometres (286,000 square miles) of Amazonian land, while 41 per cent of Colombia's rainforest area is recognized as Indian territory.

The rights of Indians to manage their traditional lands are increasingly recognized, but tragic situations have sometimes resulted when these forest cultures have come face to face with the modern world outside.

The Yanomami are the largest surviving traditional Amerindian group in the Amazon rainforest. They live in a relatively isolated upland area among the headwaters of the Orinoco and the northern tributaries of the Amazon. Over 20,000 of these people continue to live by a combination of shifting cultivation and reliance on wild forest resources. Their staple crop is the plantain, which is cultivated along with around 60 other crops that grow with careful tending even on poor soils. The Yanomami are also skilful hunters. Traditionally, they use bows made from the strong flexible wood of palm trees and arrows made from wild or cultivated canes. Different points are attached to the arrows to catch different types of prey. Barbed heads are used for shooting birds such as guans, tinamous and curassows; sharp palm-wood points, coated with poison, for hunting monkeys; and long, lanceolate heads for large land mammals, such as deer, armadillos, Capybara and peccaries. The most important prey species of all is the White-lipped Peccary.

The Yanomami have lost land to road building and mining and have suffered dreadfully from violence and introduced diseases. However, recent years have seen real progress in protecting their lands. In Brazil, 90,000 square kilometres (34,700 square miles) of forest has been recognized as belonging to the Yanomami and in Venezuela, the whole of the upper Orinoco, an area of over 80,000 square kilometres (30,000 square miles) inhabited by Yanomami and Yekuana Indians, has been declared a Biosphere Reserve. The Indians who live within the reserve have legal rights to use forest resources and are involved in decisions about managing the area.

Manioc is the staple crop of many of the Amazonian Indians. Originally native to South American forests, this tuberous plant has been cultivated in South America and the Caribbean for thousands of years. Following European settlement, the cultivation of manioc spread to tropical Africa

OPPOSITE LEFT Toasting manioc flour in Jaraua village. Manioc has to be carefully prepared to remove toxic substances.

LEFT Yanomami Indian woman. The Yanomami live in northern Brazil and southern Venezuela. Their first contacts with the outside world were only 50 years ago and they still retain much of their traditional culture.

and parts of South East Asia. In the Amazon, it is grown in garden plots cleared within the forest. It grows relatively well on the infertile forest soils, unlike crops such as maize. Clearing the forest is considered to be men's work. Nowadays, steel machetes have replaced the stone axes or piranha mandibles that were formerly used for this laborious task. Once the vegetation has been cleared, the site is burned to make use of the nutrients in the ash and then planting takes place.

The Amazonian Indians produce bread (known as *cassava*), a type of toasted flour (called *farinha*) and beers (known as *chicha*) from manioc. Grown and processed by women, it is eaten at every meal and preparing it takes up many hours of the day.

The Kayapo Indians live along the Xingu River in Brazil in a way that has hardly changed for centuries. Their deep knowledge of the wildlife of the rainforest enables them to harvest the fruits of over 250 different plants and collect hundreds of others for their tubers or leaves. The Kayapo are also skilled farmers. They clear temporary fields in the forest to grow crops, using methods that help to conserve the nutrients in the fragile soil. Occasionally a large tree is felled in order to obtain honey from a bees' nest high up in the canopy. The gap left by the felled tree is then used to grow medicinal plants.

Recently, the Kayapo people's way of life has been threatened by illegal gold prospectors and by plans for massive dams on the Xingu River. The tribe, traditionally one of the fiercest of the Amazon, is now itself fighting for survival.

Chapter Nine

Brazil

Atlantic Forests

Anyone fortunate enough to have caught sight of a
Golden Lion Tamarin in its natural habitat will know that the
experience is unforgettable. This delicate and beautifully coloured
monkey is a poignant symbol of the conservation importance,
and the fragility, of the Atlantic forests of Brazil.

Not far from the exciting carnival city of Rio, the Golden Lion
Tamarin survives, in protected fragments of forest, in one of the hottest
of the world's biodiversity hotspots. Surrounded as they are by land now
devoted to cattle ranching, these forests have a vulnerability that is all too
obvious. Yet miraculously – along with the Maned Sloth, the Seven
coloured Tanager, strange tree-dwelling cacti and countless other
wonders of the Earth – the Golden Lion Tamarin is one of many unique
and flamboyant animals and plants that, somehow, cling precariously
to existence in the rich remnant forests here.

A Narrow Belt

The Atlantic rainforests of Brazil form a narrow belt running parallel to the coast from the state of Rio Grande do Norte to Rio Grande do Sul. These forests are quite distinct from the forests of Amazonia, from which they are separated mainly by the *Cerrado*, a vast area of savannah, woodland and dry forest.

These Atlantic forests have declined dramatically in area, particularly over the past 100 years and are now reduced to fragments that amount to less than five per cent of their former range. The remnant forests are immensely important for conservation and the region is internationally recognized as outstandingly important in terms of global biodiversity. Some 20,000 plant species occur here, roughly one-third of which are restricted to this area. About half the tree species are thought to be endemic. There are over 2,000 species of butterfly (one-eighth of all the world's butterflies), of which 913 are endemic. The forests are also rich in birds; of 620 species, 73 are found nowhere else in the world. Mammal

diversity is also of great importance: 160 species out of the known total of 261 are restricted to these Atlantic forests. Sadly, but not surprisingly, much of the region's special wildlife is threatened with extinction as the forests continue to be destroyed.

Early settlers: sugar and gold

The coastal areas of Brazil were the first to be settled by European colonists, some 500 years ago. The pristine forests that once grew in the more fertile areas at the ocean's edge soon began to be cleared for agriculture. Sugarcane was grown from the early 16th century and sugar was shipped

PREVIOUS PAGE Angelica-do-Brejo trees in the Atlantic rainforest of Parana State, Southern Brazil.

LEFT Toco Toucan (*Ramphastos toco*). The name toucan is from tucano in the language of the Brazilian Tupi Indians. These wonderful rainforest birds are totally arboreal, nesting in trees and feeding high in the canopy.

ABOVE A bromeliad (*Aechmea lingulata*) of the Atlantic rainforest. Many species of *Aechmea* are cultivated as ornamental plants.

RIGHT Morpho butterflies live in the rainforests of Central and South America and are found mainly in the forest canopy. The metallic blue coloration has drawn attention to these butterflies (although the females are very drab in comparison). Males have been collected in vast quantities to make jewellery.

from Pernambuco to Portugal as early as 1526. Bit by bit, as sugarcane plantations developed along the coast of northeastern Brazil, nearly all this species-rich forest was destroyed. Wood was used to fuel the furnaces used in sugar production. Between 1600 and 1700, Brazilian sugar dominated the world market. Forests were wiped out not only for sugar, but also to provide cattle pasture. Sugarcane and cattle were the most important forms of agriculture until the early 19th century, when coffee plantations became important. Other cash crops – cocoa, rubber, oil palm – have each had a share in the destruction of the forest ecosystem.

Brazil's coastal forests have also been lost to mining and industrial development. A gold rush began at the end of the

17th century, when expeditions of *bandeirantes* from Sao Paulo discovered gold while hunting Indians. Eager to find new wealth, 300,000 Portuguese emigrated to Brazil – so great an exodus that Portugal introduced restrictions to stop people leaving the country. In the ensuing frenzy for gold, great swathes of virgin rainforest were burnt to the ground.

Too Late for the Trees?

Not surprisingly, five centuries of harvesting certain economically important tree species has also taken its toll on the Atlantic forests. Pau Brasil, the tree that gave its name to Brazil, has been traded internationally since the earliest times of European settlement. Initially, it was exploited as a very valuable source of red dye. With the 19th-century introduction of synthetic dyes, the trade in dyewood collapsed. Pau Brasil remains a valuable commodity, however, as it is the main wood used to make violin bows. Today, after centuries of exploitation, Pau Brasil – an endemic species of the Atlantic rainforest – faces extinction in the wild.

In 1997, a workshop was held in Brazil to draw up a plan of action to save Pau Brasil. Fauna & Flora International helped to organize the workshop as part of its Soundwood Programme, aimed at conserving tree species that are used to make musical instruments. The meeting brought together botanists, foresters, timber traders and musical-instrument manufacturers, all of whom have a long-term interest in the survival of tree species. The workshop resulted in an ambitious conservation plan. The steps taken to implement it have so far focused on education, but much more remains to be done.

Another endemic tree that has been heavily exploited for its timber is the vulnerable Brazilian Rosewood. One of the most highly prized woods in Brazil, the timber of the rosewood has been harvested since colonial times for use in decorative veneers, high-quality furniture as well as musical instruments. The Brazilian Rosewood also yields useful resin and oil and is attractive as an ornamental tree. Although there is some small-scale cultivation, no plantations have been developed to meet the international demand for this valuable timber. The highest remaining concentrations of the species are found in forest on rich soils in southern Bahia and northern Espírito Santo – areas where rates of deforestation are high. In recent years, with most rosewood trees already logged, timber cruisers, known as *madereiros*, have painstakingly searched large areas to find any remaining stands of the trees, which they can cut and sell for extremely high prices.

Fortunately, Brazilian Rosewood is found in a number of protected areas and is included on the official list of threatened Brazilian plants. The export of its logs has been banned for over 30 years and in 1992 the species was added to Appendix I of CITES. This outlaws international trade in timber taken from the wild. Sadly, despite these protective measures, illegal cutting and timber smuggling are putting the survival of the species at risk.

LEFT An Emerald Tree Frog in the cloud forest of Sierra dos Orgaos National Park, Brazil. This scenically stunning national park in Rio de Janeiro State also protects an important diversity of plant species.

RIGHT This Jararaca pitviper is a fairly common species of the Brazilian Atlantic rainforest.

BELOW Green-winged Macaws and Scarlet Macaws feed on soil at a riverbank in Manu National Park, Peru. This magnificent national park is a popular destination for nature lovers from around the world.

Primates on the Brink

Animals in serious trouble in the Atlantic forests include all four species of lion tamarin. Only around 600 individuals of the critically endangered Golden Lion Tamarin remain in the wild. They live in forest remnants in the northern part of the state of Rio de Janeiro. Even fewer individuals of the critically endangered Black-faced Lion Tamarin currently survive. This species was not discovered until 1990, in the state of Parana.

The Black Lion Tamarin, the third critically endangered species, was thought to be extinct until it was found again in 1970. It is now known to occur in five forest patches in the state of Sao Paulo, with a total population of less than 1,000

individuals. The Golden-headed Lion Tamarin is the most numerous of the four species, but even though between 6,000 and 15,000 individuals remain, it is classified as endangered.

Measures to conserve the tamarins in their natural habitat include the creation of protected areas, such as the Poco das Antas Biological Reserve for the Golden Lion Tamarin. These efforts are reinforced by captive breeding programmes, which have been particularly successful for the Golden Lion Tamarin and the Golden-headed Lion Tamarin and have allowed some to be returned to the wild. The Black-faced Lion Tamarin is not yet being bred in captivity, but may well be soon.

Like the tamarins, the two species of muriqui or woolly spider monkey are also endemic to the Atlantic forests, where they are seriously threatened. The muriqui, largest of the South American monkeys, has been hunted to the brink of extinction. Early European explorers found such large muriqui populations that they were able to use them as their main source of meat. Today, combined numbers of the two recognized species, *Brachyteles hypoxanthus* and *B. arachnoides*, number no more than about 1,150 individuals. Hunting is still a threat but, as is so often the case, an even greater worry is the relentless destruction of the animals' habitat.

LEFT The Brazilian Three-toed Sloth spends most of its life in trees. Green algae frequently grow on its hairy coat, helping it to blend into the forest background. Protected by law from hunting, habitat destruction is now the main threat to this strange creature.

ABOVE The Golden Lion Tamarin is a delightful primate that fortunately has benefited from successful international conservation efforts.

RIGHT Squirrel monkeys are found in primary and secondary forests, especially along streams. They form larger groups than any other New World monkeys: it has been reported that a troop of 550 were found in the Amazon.

Chapter Ten

Temperate Rainforests

Cathedrals of the Natural World

In the temperate rainforests, widely scattered across
the world from Norway to New Zealand, the steamy, humid
atmosphere of the tropical rainforests is replaced by a cool tranquillity.
The temperate forests may not match those of the tropics in terms of
biodiversity, but they engender the same awe-inspiring sense of natural
wonder. Nourished by the abundant rainfall that characterizes many
coastal regions in the temperate zones, the huge ancient trees of these
damp and misty forests include the tallest living organisms in the world.
It is hard to accept that the future of such giants is by no means secure.
Yet, in many places, temperate rainforests are suffering the same fate as
their tropical counterparts, and are being felled for economic gain and in
the interests of so-called progress.

The Global Picture

Classifying the world's forests is an imprecise science, and many different categories and definitions exist. Scientists have recognized temperate rainforests as a distinct forest type for the past 50 years. The boundaries between these forests and the subtropical and tropical rainforests with which they merge in continental areas are blurred as the species make-up of one type gradually changes into that of another.

Temperate broad-leaved rainforests occur in places that could scarcely be more widely scattered across the world. They are found in parts of South East Asia, China, Korea and southern Japan, eastern Australia and southern Brazil. Broad-leaved evergreen rainforests, with conifers in the north, also cover much of New Zealand. In the Southern Hemisphere, Chile has the largest area of coastal temperate rainforest. The rich broad-leaved forest in the south of the country is known to

scientists as Valdivian rainforest. Tree communities including the majestic *Araucaria* and *Fitzroya* conifers intermingle in these rich rainforests of the Southern Hemisphere.

In the southeastern United States, patches of warm temperate rainforest are found near the coast from Louisiana, Florida and Georgia to North Carolina. Evergreen oaks and the Evergreen Magnolia are the dominant trees of this ancient and now very restricted forest type. Far more familiar

PREVIOUS PAGE Native podocarp forest in Pureora Forest Park, New Zealand. The podocarp and mixed forests on the western slopes of the Southern Alps form one of the world's most extensive temperate rainforests.

LEFT Karamea Forest, west coast, South Island, New Zealand – a place of dramatic contrasts.

RIGHT North Island Kaka – this species has suffered from deforestation and competition with introduced species. It is no longer found on mainland New Zealand.

are the imposing coastal rainforests of the North American Pacific seaboard, known for their colossal trees. Half of all the world's coastal rainforests are found in the strip that runs for much of the length of this Pacific coast of North America.

Small fragments of temperate rainforest remain in Norway, and some of Japan's forests may also fall within this category. Temperate coastal rainforest also occurs on the west coast of New Zealand's South Island and in western Tasmania. The temperate rainforest that is least well known worldwide is probably the unique form that occurs around the Black Sea area of Transcaucasia.

Temperate coastal rainforest has only very recently come to be recognized as a distinct ecosystem type. Worldwide, coastal

rainforests comprise only two to three per cent of temperate forests. They are distinguished by their closeness to the sea, and by the presence of coastal mountains. As with all rainforests, they can grow only in places where rainfall is high throughout the year.

Although they are widely scattered geographically, the temperate coastal rainforests share remarkable similarities. They are dominated by large, long-lived trees. They differ from tropical rainforests in not having such a great diversity of species, but they are extraordinarily productive, accumulating more organic matter than any other forest ecosystem on Earth. These majestic rainforests are immensely important to conservation.

LEFT Giant Tree Frog. The moist air within rainforests allows thin-skinned tree frogs to live amongst the trees.

OPPOSITE LEFT In Tasmania, the Ringtail Possum has a thick pelt to cope with the cold winters. This has made the possum very attractive to hunters for the fur trade. The Ringtail Possum is a tree-living species that feeds at night on leaves.

OPPOSITE RIGHT Swainson's Lorikeet (otherwise known as Rainbow Lorikeet). Vividly coloured and noisy, this bird chatters and screeches in large flocks as it feeds.

Tasmania

Australia's largest tracts of cool temperate rainforest cover around 10 per cent of Tasmania, in areas where the annual rainfall generally amounts to more than 1,200 millimetres (47 inches). The largest area of rainforest is in the northwest of the island, though it is also found elsewhere in the west, and in patches in the northeastern highlands. Tiny areas also survive in some east-coast gullies, where moisture from clouds or streams compensates for the slightly lower levels of rainfall.

Tasmanian rainforest has been described as having a cathedral-like quality. The majestic trees provide a cool, dark, damp habitat, with abundant cover of mosses and lichens. In autumn and early winter, the forest floor comes to life with an explosion of brightly coloured fungi.

Plant life

Tasmanian rainforest is dominated by species of *Nothofagus*, especially Myrtle Beech, which reaches a height of up to 15 metres (50 feet). Other important rainforest trees include Sassafras, which commonly grows in association with myrtle, the endemic Leatherwood, Celery-top Pine, King Billy Pine,

Huon Pine, and Chestnut Pine. These trees are highly valued by the craft wood industry. Unfortunately, because they are slow-growing, it is not economic to cultivate them in plantations, and the future long-term supply is uncertain. At present, there are export embargoes on Huon and King Billy Pine. The salvage of dead Huon Pine logs from hydroelectric sites supplies much of the current demand, but the long-term supply of timber from this ancient species is far from secure.

The native tree fern *Dicksonia antarctica*, which flourishes in the shade of the huge trees, provided a food source for the Tasmanian aboriginal people. This stately plant is one of the few tree ferns that can survive outdoors in the Northern Hemisphere, and has, therefore, recently become popular in European gardens. The ferns for export are harvested from the wild, in accordance with approved management plans.

Tasmanian rainforests are generally divided into four main types – callidendrous (tall trees), thamnic (shrubby), implicate (tangled), and montane – each with a different composition of species. The variation is mainly due to differences in soil, rainfall, aspect and altitude. There is also an intermediate forest type where rainforest species grow beneath tall eucalypts.

Special creatures

Tasmanian rainforest is such a tranquil environment that it may appear to harbour few animals. It is true that vertebrates are both less diverse and less abundant than in some other types of forest, but Tasmania's forests are home to many very special creatures. Their names alone often have a special fascination. Mammals include the Tasmanian Long-tailed Mouse, the Ringtail Possum, the Pademelon, the Spotted-tailed Quoll and the Dusky Antechinus. Over 20 species of native bird regularly visit the rainforest, including Black Currawong, Green Rosella, Olive Whistler and Grey Goshawk. The Tasmanian Tree Frog lives here, and common reptiles include Tiger Snake and Brown Skink. Tasmania's rainforest contains some of the most ancient and primitive invertebrates, including the large land snail, Macleay's Swallowtail Butterfly, freshwater crayfish and the Peripatus, or Velvet Worm.

Safeguarding the forests

During the 1980s, a major environmental battle was fought to save the rainforests of Tasmania. The cause of the conflict was a plan agreed by the Tasmanian Government in 1982 to dam the lower reaches of the Gordon and Franklin Rivers as part of a major hydroelectricity scheme. The Wilderness Society of Australia led the protests and organized a major blockade of the proposed dam site. Over 1,000 people were arrested during the protest campaign. The protesters' efforts were rewarded in July 1983. The High Court of Australia ruled against the building of the dam and construction work was stopped.

Part of the Tasmanian rainforest had already been declared a World Heritage site in 1982. Four years later, logging in adjacent areas led to further protests and, after a Commission of Enquiry, the World Heritage site was extended. Further areas have subsequently been added and the World Heritage site now comprises four national parks, three state reserves, two Protected Archaeological Sites and a range of other land designations. In total the protected area covers some 1.4 million hectares (3.5 million acres).

The greatest danger to Tasmania's rainforest is fire, both deliberate and accidental. In the last century, over seven per cent of the rainforest in Tasmania was burnt. If left undisturbed, burnt forest eventually recovers, but this process may take several hundred years. More usually, the land is converted to other uses such as agriculture, forest plantations, dams and mining. Another threat is from pests and diseases. Myrtle wilt, for example, is a serious fungal disease that kills myrtles, especially in cases where there has been some form of disturbance. Despite the threats, conservation is well established.

Visitors to the rainforests of Tasmania are well catered for, and there are excellent trails and walks through the national parks and at Liffey Falls on the northern edge of the Tasmanian Wilderness World Heritage Area.

North America

Coastal rainforest extends for some 3,000 kilometres (2,000 miles) along the Pacific coast of North America from northern California to Alaska. The strip of coastal land in between includes the rich forests of the Olympic Peninsula and the Canadian coast, together with the Sitka Spruce forests of southeastern Alaska and adjacent Canada. Towards the southern end of this long strip of rainforest are the legendary Coast Redwood forests of northern California and southern Oregon.

North America's coastal forests contain about 30 tree species, and a combined total of about 250 mammal and bird species. Certain of the birds and fish are uniquely adapted to a life that depends on the sea and the rainforests in equal measures. The seven species of Pacific salmon and trout, for example, have a wide range in the Pacific Ocean, but return to their birthplace in the rainforest streams to spawn. The Marbled Murrelet, a small lark-sized seabird, spends its days feeding at sea, but returns at night to its nest on the mossy upper branches of old-growth trees. It is classified as a threatened species under US legislation. The main reason for its decline has been logging, for timber harvesting has been concentrated along the coast, in prime murrelet habitat. The murrelet is also highly susceptible to oil spills, since it spends most of its time on the sea surface, feeding in local concentrations close to the shore. The *Exxon Valdez* spill in 1989 killed more than 600 murrelets.

Perhaps the most famous symbol of the North American rainforests is the Northern Spotted Owl, which has been the cause of intense battles between conservationists, loggers and local politicians for the past 25 years. The total population of this owl is estimated at 3,000-5,000 pairs, in a range that extends from southwestern British Columbia to northern California. The main threat faced by this owl is, again, logging. After acrimonious debate, the Northern Spotted Owl was listed as threatened under the US Endangered Species act in 1990. An area of nearly three million hectares (seven million acres) of federal land was subsequently declared critical habitat for the species under the terms of the legislation, but the controversy over logging on government land continues.

The Coast Redwood is a surviving representative of the forests that covered large tracts of the Northern Hemisphere 140 million years ago. These slim and elegant tree giants grow to over 91 metres (300 feet) in height. The tallest tree currently recorded in the world is a Coast Redwood, aptly named Tall Tree, in Redwood National Park in northern California. It measures more than 110 metres (360 feet) tall, and is estimated to be over 1,500 years old.

For centuries, North America's coastal rainforests have sustained native people, living simple lives in harmony with their environment. The Sinkyone people, who no longer survive, believed that the Coast Redwoods were sacred protectors of the forests and streams, and guardians of their ancestral spirits. Early European settlers and fur traders founded settlements in the forested valleys and along the coast. Their livelihoods depended on fishing, hunting, and forest produce – still vital to their descendants in some areas today. More than 60 different languages were spoken by the native people of the rainforests, but sadly more than 50 of them are now lost.

OPPOSITE LEFT The Spotted Owl has become a flagship species for the conservation of the magnificent temperate rainforests of North America. Powerful logging interests remain a threat to the habitat of this species.

OPPOSITE RIGHT The Banana Slug can grow to a length of 20 centimetres (8 inches), with occasional giants reaching 25.4 centimetres (10 inches), and can weigh a quarter of a pound (0.1 kg).

ABOVE Moss and lichen-covered trees in the Olympic National Park, Washington. This national park is easy to reach by ferry and road from Seattle. There are good facilities for visitors, including the Hoh Rain Forest Visitor Centre.

Alaska

The coastal rainforest of Alaska is probably the largest area of pristine temperate rainforest in the world. Dominated by hemlock and spruce, the forest here is a very diverse habitat and contains important populations of flagship species such as the Grizzly Bear and the Bald Eagle.

The Tongass National Forest covers about 6.8 million hectares (17 million acres) of coastal forest in southeastern Alaska. It is the largest national forest in the USA. Only about 57 per cent of the land is actually forested, the rest being mainly mountain and tundra. The USA's second largest forest system is also in Alaska: Chugach National Forest covers around 2.4 million hectares (six million acres), in three biogeographical regions: the Kenai peninsula, Prince William Sound, and the Copper and Bering Delta region.

Although large areas of land are protected in national or state parks, the old-growth forests of Alaska are rapidly disappearing. The main pressure is from the logging industry, particularly to supply Japan, the destination for over 70 per cent of the total exports. Demand is mainly for raw logs.

The exploitation of Tongass National Forest has become a controversial conservation issue. Legislation in 1947 allowed the signing of long-term timber contracts involving national forest land, despite outstanding claims to some of the area by native people. Timber extraction has been dominated by two large companies, which benefit from very favourable 50-year leases. These logging terms were introduced to encourage trade, and to discourage Japan from buying its timber from Russia. Logging in Alaska is heavily subsidized by the State, and would otherwise not be profitable. In 1990, the Tongass Timber Reform Act was introduced. This increased protected area coverage and placed some restrictions on federal support for the timber industry. In spite of this, 1991 saw the highest level of logging for a decade. The Kenai peninsula and Prince William Sound are also subject to logging.

About 16 per cent of North America's coastal rainforests is protected, mostly in Alaska. Other outstanding protected areas of the temperate rainforest here include the Olympic National Park, in Washington state near Seattle. This was established in 1938 and has subsequently been given World Heritage status and designated a Biosphere Reserve. According to the IUCN, Olympic National Park is the best natural area in the entire Pacific Northwest, with a spectacular coastline, scenic lakes, majestic mountains and glaciers, and magnificent temperate rainforests; these are outstanding examples of ongoing evolution and superlative natural phenomena. There is nowhere quite like it in the world.

OPPOSITE LEFT Club moss and licorice ferns growing on Big Leaf Maples of Olympic National Park's temperate rainforest. Sitka Spruce and Western Hemlock are the dominant trees of the rainforest, with Big Leaf Maple, Douglas Fir, Red Alder and Black Cottonwood also found throughout the forest.

ABOVE Big Leaf Maples are deciduous trees with a bark which retains moisture. In the Pacific Northwest its trunk and large branches are often covered with mosses, liverworts and ferns, such as the sword fern, with its tapering, green fronds.

LEFT Pacific salmon, such as this Sockeye Salmon, and the northern rainforests are interdependent. During the upstream migration, the remains of salmon caught as prey add an important source of nitrogen to the forest ecosystem.

149

Original Rainforest Areas of the World

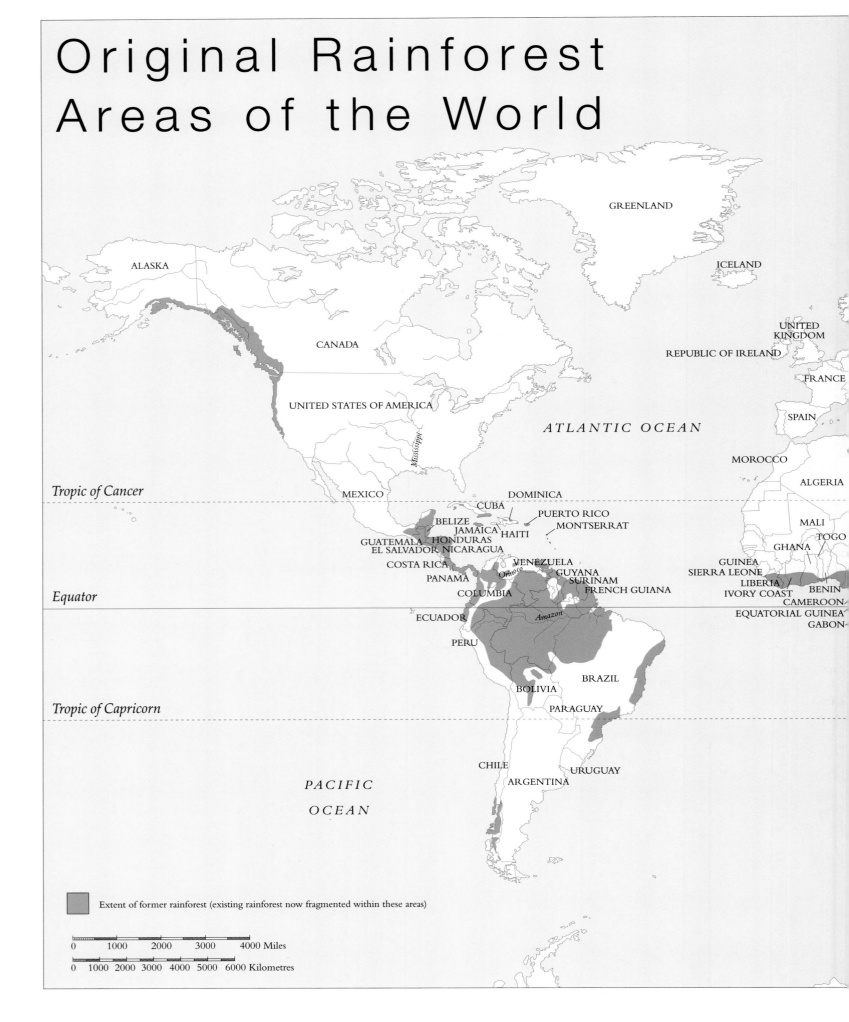

GREENLAND

ALASKA

ICELAND

CANADA

UNITED KINGDOM

REPUBLIC OF IRELAND

FRANCE

UNITED STATES OF AMERICA

ATLANTIC OCEAN

SPAIN

Mississippi

MOROCCO

ALGERIA

Tropic of Cancer

MEXICO

DOMINICA

CUBA

PUERTO RICO

MALI

BELIZE

MONTSERRAT

JAMAICA

HAITI

TOGO

GUATEMALA
EL SALVADOR

HONDURAS
NICARAGUA

GHANA

COSTA RICA

VENEZUELA

GUINEA
SIERRA LEONE

PANAMA

Orinoco

GUYANA

LIBERIA

BENIN

COLUMBIA

SURINAM
FRENCH GUIANA

IVORY COAST

CAMEROON

Equator

ECUADOR

Amazon

EQUATORIAL GUINEA

GABON

PERU

BRAZIL

BOLIVIA

Tropic of Capricorn

PARAGUAY

CHILE

URUGUAY

PACIFIC

ARGENTINA

OCEAN

Extent of former rainforest (existing rainforest now fragmented within these areas)

0   1000   2000   3000   4000 Miles

0   1000 2000 3000 4000 5000 6000 Kilometres

150

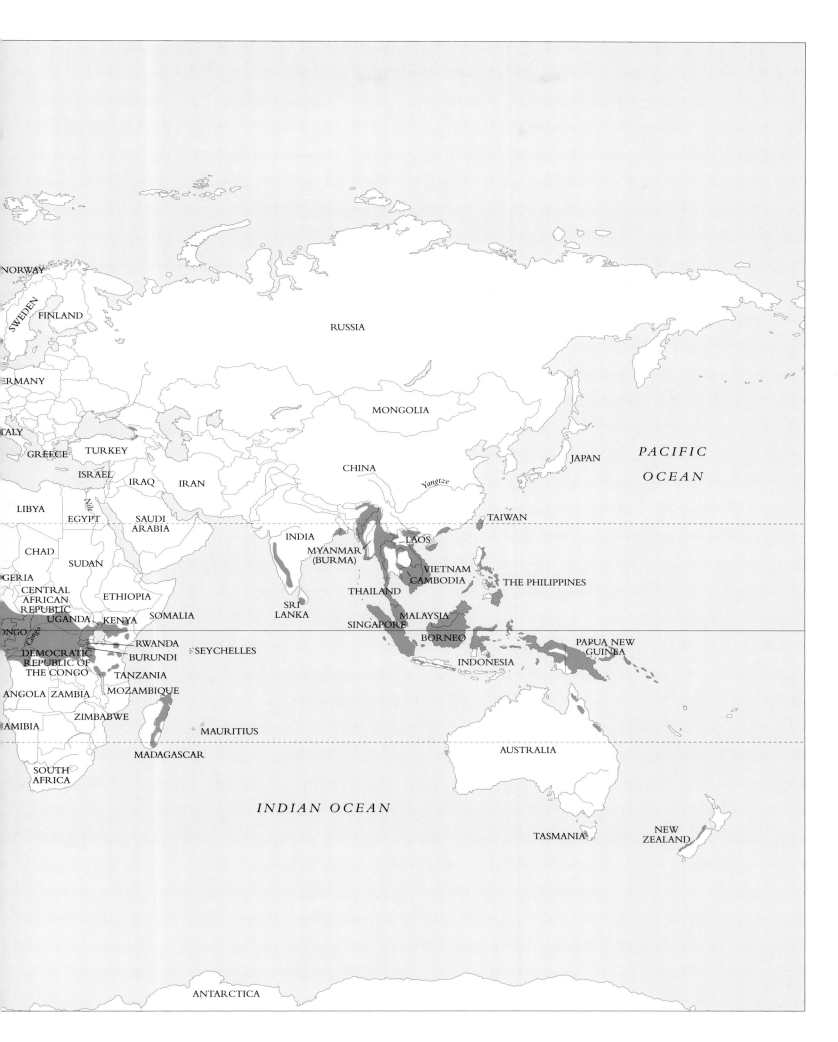

NORWAY

SWEDEN FINLAND

RUSSIA

ERMANY

TALY

GREECE TURKEY

MONGOLIA

JAPAN

PACIFIC

OCEAN

ISRAEL

IRAQ IRAN

CHINA

Yangtze

LIBYA

EGYPT

SAUDI
ARABIA

Nile

TAIWAN

INDIA

CHAD

SUDAN

MYANMAR
(BURMA)

LAOS

VIETNAM
CAMBODIA

THE PHILIPPINES

GERIA

CENTRAL
AFRICAN
REPUBLIC

ETHIOPIA

THAILAND

SRI
LANKA

UGANDA KENYA

SOMALIA

MALAYSIA

ONGO

Congo

SINGAPORE

BORNEO

PAPUA NEW
GUINEA

RWANDA

BURUNDI

SEYCHELLES

INDONESIA

DEMOCRATIC
REPUBLIC
OF THE CONGO

TANZANIA

ANGOLA ZAMBIA

MOZAMBIQUE

AMIBIA

ZIMBABWE

MAURITIUS

MADAGASCAR

AUSTRALIA

SOUTH
AFRICA

INDIAN OCEAN

TASMANIA

NEW
ZEALAND

ANTARCTICA

Rainforest Species of Concern

Mammals and birds have been well studied around the world and there is good information on their conservation status. For other animal groups, such as reptiles, amphibians, fishes and invertebrates, there is often little agreement on how many species there are never mind their conservation status and reliance on forests. For example, there are thought to be 7,900 species of reptiles, but the conservation status of over 85 per cent of these has yet to be assessed. For invertebrates, a group that includes 95 per cent of all animal species, the situation is even worse. In total there may be around 950,000 species of insects, but the status of 99.94 per cent of the species has not been assessed and no-one knows how many occur in rainforests. For these poorly studied groups, the degree of threat and the importance of forests to their long-term survival may therefore currently appear small, but from the limited information we have these factors are grossly underestimated.

Conservation status information is reasonably good for trees, for at least some rainforest areas, but for most other plant groups information is very patchy. As with invertebrates, the rainforest mosses, liverworts and fungi are scarcely studied.

The species listed below are examples of rainforest species of concern, many of which are also referred to in the main text of the book. For each species the global conservation status, according to IUCN – The World Conservation Union, and geographical distribution is given.

Mammals

There are over 4,700 species of mammals, and over 1100 are considered threatened. Of these, 33 per cent occur in lowland rain forest and 22 per cent in montane rain forest.

Threatened primates

- Golden Lion Tamarin *Leontopithecus rosalia*, Critically Endangered, Brazil
- Black-faced Lion Tamarin *Leontopithecus caissara*, Critically Endangered, Brazil
- Golden-headed Lion Tamarin *Leontopithecus chrysomela*, Endangered, Brazil
- Woolly Spider Monkey *Brachyteles arachnoides*, Critically Endangered, Brazil
- Bald-headed Uacari *Cacajao calvus*, Vulnerable, Brazil, Colombia and Peru
- Gorilla *Gorilla gorilla*, Endangered, Central and West Africa
- Bonobo or Pygmy Chimpanzee *Pan paniscus*, Endangered, Democratic Republic of Congo and possibly Congo
- Drill *Mandrillus leucophaeus*, Endangered, West Africa
- Golden Lemur *Hapalemur aureus*, Critically Endangered, Madagascar
- Aye-aye *Daubentonia madagascariensis*, Endangered, Madagascar
- Golden-brown Mouse Lemur *Microcebus ravelobensis*, Endangered, Madagascar
- Indri *Indri indri*, Endangered, Madagascar
- Hairy-eared Dwarf Lemur *Allocebus trichotis*, Endangered, Madagascar
- Red-bellied Lemur *Eulemur rubriventer*, Vulnerable, Madagascar
- Tonkin Snub-nosed Monkey *Rhinopithecus avunculus*, Critically Endangered, Vietnam
- Francois's Langur *Trachypithecus francoisi*, Vulnerable, China, Laos and Vietnam
- Long-nosed or Proboscis Monkey *Nasalis larvatus*, Vulnerable, Brunei, Indonesia and Malaysia
- Black Crested Gibbon *Nomascus concolor*, Endangered, Cambodia, China, Laos and Vietnam
- Orang-utan *Pongo pygmaeus*, Endangered, Malaysia, Indonesia and possibly Brunei.

Threatened elephants, rhinos, hippos and small mammals

- African Elephant *Loxodonta africana*, Endangered, sub-Saharan Africa
- Asian Elephant *Elephas maximus*, Endangered, South and South East Asia
- Javan Rhinoceros *Rhinoceros sondaicus*, Critically Endangered, South East Asia
- Sumatran Rhinoceros *Dicerorhinus sumatrensis*, Critically Endangered, South East Asia
- Pygmy Hippopotamus *Hexaprotodon liberiensis*, Vulnerable, West Africa
- Tamaraw *Bubalus mindorensis*, Critically Endangered, Philippines
- Asian Tapir *Tapirus indicus*, Vulnerable, South East Asia
- Baird's Tapir *Tapirus bairdii*, Vulnerable, Latin America
- Jentink's Duiker *Cephalophus jentinki*, Vulnerable, West Africa
- Almiqui or Cuban Solenodon *Solenodon cubanus*, Endangered, Cuba
- Spotted-tailed Quoll *Dasyurus maculatus*, Vulnerable, Australia

Threatened carnivores

- Clouded Leopard *Neofelis nebulosa*, Vulnerable, Asia
- Jaguar *Panthera onca*, Near Threatened, Latin America
- Liberian Mongoose *Liberiictis kuhni*, Endangered, Cote d'Ivoire, Guinea, Liberia
- Owston's Palm Civet *Chrotogale owstoni*, Vulnerable, China, Laos and Vietnam
- Malayan Sun bear *Helarctos malayanus*, Data Deficient, South East Asia
- Small-eared Dog *Dusiyon microtis*, Data Deficient, South America

Threatened bats

Nearly a quarter of all mammal species are bats and many of these are dependent on forests for feeding and roosting. Almost a quarter of all bat species are threatened.

- Philippines Tube-nosed Fruit Bat *Nyctimene rabori*, Critically Endangered, Philippines
- Malay Leaf-nosed Bat *Hipposideros nequam*, Critically Endangered, Malaysia
- Flat-headed Myotis *Myotis planiceps*, Critically Endangered, Mexico
- Thomas's Big-eared Bat *Pharotis imogene*, Critically Endangered, New Guinea.
- Bumblebee Bat *Craseonycteris thonglongyai*, Endangered, Thailand

Threatened aquatic mammals

- Giant Otter *Pteroneura brasiliensis*, Endangered, South America
- Amazonian Manatee *Trichechus inunguis*, Vulnerable, South America
- Amazon River Dolphin *Inia geoffrensis*, Vulnerable, South America

Birds

There are almost 10,000 species of birds worldwide. Over 1,100 species are threatened and almost 900 are entirely dependent on forests. Over 90 per cent of threatened forest birds occur only in the tropics and over 80 per cent occur in moist forests. Lowland moist forests contain 41 per cent of the threatened bird species and montane moist forests 36 per cent.

- Philippine Cockatoo *Cacatua haematuropygia*, Critically Endangered, Philippines
- Cebu Flowerpecker *Dicaeum quadricolor*, Critically Endangered, Philippines
- Philippine Eagle *Pithecophaga jefferyi*, Critically Endangered, Philippines
- Philippine Hawk Eagle *Spizaetus philippinensis*, Vulnerable, Pilippines
- Maleo *Macrocephalon maleo*, Vulnerable, Indonesia
- Clarke's Weaver *Ploceus golandi*, Endangered, Kenya
- Usambara Eagle Owl *Bubo vosseleri*, Vulnerable, Tanzania
- Sokoke Scops Owl *Otus ireniae*, Endangered, Kenya, Tanzania
- Congo Peafowl *Afropavo congensis*, Vulnerable Democratic Republic of Congo
- Magdalena Tinamou *Crypturellus saltuarius*, Critically Endangered, Colombia
- Kalinowski's Tinamou *Northoprocta kalinowskii*, Critically Endangered, Peru
- Black Tinamou *Tinamus osgoodi*, Vulnerable, Colombia and Peru
- Choco Tinamou *Crypturellus kerriae*, Vulnerable, Panama and Colombia
- Taczanowski's Tinamou *Northoprocta taczanowskii*, Vulnerable, Peru.
- Horned Guan *Oreophasis derbianus*, Endangered, Mexico and Guatemala
- Bare-necked Umbrellabird *Cephalopterus glabricollis*, Vulnerable, Costa Rica and Panama
- Long-wattled Umbrellabird *Cephalopterus penduliger*, Vulnerable, Colombia and Ecuador
- Three-wattled Bellbird *Procnias tricarunculata*, Vulnerable, Costa Rica, Panama, Nicaragua and Honduras
- Military Macaw *Ara militaris*, Vulnerable, Central America and South America
- Golden Parakeet *Guaruba guarouba*, Endangered, Brazil
- Montserrat Oriole *Icterus oberi*, Critically Endangered, Montserrat
- Ivory-billed Woodpecker *Camphephilus principalis*, Critically Endangered, Cuba
- Cuba Giant Kingbird *Tyrannus cubensis*, Endangered, Cuba
- Marbled Murrelet *Brachyramphus marmoratus*, Vulnerable, Canada and USA
- Spotted Owl *Strix occidentalis*, Near Threatened, Canada, USA and Mexico

Trees

There are an estimated 80,000 - 100,000 tree species worldwide with at least 50,000 species in tropical forests. Currently around 8,000 tree species are considered to be globally threatened. The list is likely to grow as more information becomes available from remote rainforest areas and as more forests are opened up for logging and agricultural settlement. The examples given here are all valuable timber species and some also provide useful medicinal products.

- Agar wood *Aquilaria malaccensis*, Vulnerable, Indian Subcontinent and South East Asia
- Monkey Puzzle Tree *Araucaria araucana*, Vulnerable, Argentina, Chile
- King Billy Pine *Athrotaxis selaginoides*, Vulnerable, Australia (Tasmania)
- Moabi *Baillonella toxisperma*, Vulnerable, Cameroon, Congo, Gabon, Nigeria
- Cedro *Cedrela fissilis*, Endangered, Central and South America
- Madagascar Rosewood *Dalbergia delphinensis* (and other species), Endangered, Madagascar
- Brazilian Rosewood *Dalbergia nigra*, Vulnerable, Brazil (Atlantic Forest)
- Sulawesi Ebony *Diospyros celebica*, Vulnerable, Indonesia (Sulawesi)
- Tiama or African Mahogany *Entandrophragma angolense*, Vulnerable, Africa
- Sapele or African Mahogany *Entandrophragma cylindricum*, Vulnerable, Africa
- Utile or African Mahogany *Entandrophragma utile*, Vulnerable, Africa
- Belian *Eusideroxylon zwageri*, Vulnerable, Brunei, Indonesia, Malaysia, Philippines
- Alerce *Fitzroya cupressoides*, Endangered, Argentina, Chile
- Lignum Vitae *Guaiacum officinale*, Endangered, Caribbean Islands, Colombia, Venezuela
- Huon Pine *Lagarostrobos franklinii*, Vulnerable, Australia (Tasmania)
- Afrormosia *Pericopsis elata*, Endangered, Cameroon, Congo, Côte d'Ivoire, DR Congo, Ghana, Nigeria
- Kuku or Nedun *Pericopsis mooniana*, Vulnerable, South East Asia
- Merkus Pine *Pinus merkusii*, Vulnerable, Indonesia, Philippines, Vietnam
- Pau Rosa *Swartzia fistuloides*, Endangered, West and Central Africa
- Mahogany *Swietenia macrophylla*, Vulnerable, Central and South America
- Cuban or Spanish Mahogany *Swietenia mahagoni*, Endangered, Caribbean islands and Florida
- Makoré *Tieghemella heckelii*, Endangered, Cameroon, Côte d'Ivoire, Gabon, Ghana, Liberia, Nigeria, Sierra Leone
- Molave *Vitex parviflora*, Vulnerable, Indonesia, Malaysia, Philippines
- *Voanioala gerardii*, Critically Endangered, Madagascar
- Mangkono *Xanthostemon verdugonianus*, Vulnerable, Philippines

Information Sources

BirdLife International, 2000. *Threatened Birds of the World*. Lynx Edicions, Barcelona and BirdLife International, Cambridge.

Groombridge, B and Jenkins, M.D., 2000. *Global Biodiversity. Earth's Living Resources in the 21st Century*. World Conservation Press, Cambridge.

Hilton-Taylor, C. compiler, 2000. *2000 IUCN Red List of Threatened Species*. IUCN, Gland, Switzerland and Cambridge, UK.

Oldfield, S.F., Lusty, C. and MacKinven, A. compilers, 1998. *The World List of Threatened Trees*. World Conservation Press, Cambridge.

UNEP-WCMC website: www.unep-wcmc.org

Useful Addresses

There are many organizations working to save the world's rainforests; to protect the rights of indigenous people who live there; and to conserve threatened rainforest plants and animals. The list below gives a selection of the key organizations. Some of these are membership organizations (marked with a ★), others are intergovernmental organizations and information centres. Also included are organizations that promote sustainable tourism.

BirdLife International★
Wellbrook Court
Girton Road
Cambridge CB3 0NA
UK
Tel: +44 1223 277318
Fax: +44 1223 277200
Email: birdlife@birdlife.org.uk

Conservation International★
1919 M Street
NW Suite 600
Washington DC 20036
USA
Tel: +1 202 916 1000
Website:www.conservation.org

Fauna & Flora International★
Great Eastern House
Tenison Road
Cambridge CB1 2DT
UK
Tel: +44 1223 571000
Fax: +44 1223 461481
Website: www.fauna-flora.org

**Friends of the Earth
International Secretariat**
P.O. Box 19199
1000 GD Amsterdam
The Netherlands
Tel.: +31 20 622 1369
Fax: +31 20 639 2181
Website: www.foei.org

**Friends of the Earth
(England, Wales, Northern
Ireland) ★**
26-28 Underwood Street
London N1 7JQ
UK
Tel: +44 207 490 1555
Fax: +44 207 490 0881
Website: www.foe.co.uk

**Friends of the Earth
Scotland★**
72 Newhaven Road
Edinburgh EH6 5QG
Scotland
UK
Tel: +44 131 554 9977
Fax: +44 131 554 8656
Website: www.foe-scotland.org.uk

Friends of the Earth US★
1025 Vermont Ave NW
3rd Floor
Washington DC 20005-6303
USA
Tel: +1 202 783 7400
Fax: +1 202 783 0444
Website: www.foe.org

Forest Stewardship Council
Avenida Hidalgo 502
68000 Oaxaca
Mexico
Tel & Fax: +51 951 62110
Website: www.fscoax.org

Global Forest Watch
World Resources Institute
10 G Street NE
Washington DC 20002
USA
Tel: +1 202 729 7600
Fax: +1 202 729 7686
Website:
www.globalforestwatch.org

Greenpeace International
Keizersgracht 176
1016 DW Amsterdam
The Netherlands
Tel: +31 20 523 6222
Fax: +31 20 523 6200
Website: www.greenpeace.org

Greenpeace UK ★
Canonbury Villas
London N1 2PN
UK
Tel: +44 207 865 8100
Fax: +44 207 865 8200

Greenpeace USA ★
702 H Street NW
Suite 300
Washington DC 20001
USA
Tel: +1 202 462 1177
Fax: +1 202 462 4507

**IUCN – The World
Conservation Union**
Rue Mauverney 28
1196 Gland
Switzerland
Website: www.iucn.org

Native Forest Network★
PO Box 301
Deloraine
Tas 7304
Australia
Tel: +61 3 6369 5102
Fax: +61 3 6369 5150
Website: www.nfn.org.au

Rainforest Action Network★
221 Pine Street Suite 500
San Francisco CA 94104
USA
Tel: +1 415 398 4404
Fax: +1 415 398 2732
Website: www.ran.org

Rainforest Alliance★
65 Bleecker Street
New York NY 10012
USA
Tel: +1 212 677 1900
Website: www.rainforest-alliance.org

Rainforest Foundation UK★
Suite A5, City Cloisters
196 Old Street
London EC1V 9FR
UK
Tel: +44 207 251 6345
Fax:+44 207 251 4969
Website:
www.rainforestfoundationuk.org

Rainforest Foundation USA★
270 Layfayette Street #1107
New York NY 10012
USA
Tel: +1 212 431 9098
Fax: +1 212 431 9197
Website: www.savetherest.org

**Rainforest Information
Centre★**
P.O Box 368
Lismore
NSW 2480
Australia
Tel: +61 266 218505
Website: www.forests.org/ric

Rainforest Rescue★
The National Arbor Day
Foundation
100 Arbor Avenue
Nebraska City NE 68410
USA
TEL: +1 402 474 5665
Website: www.arborday.org

Tourism Concern
Stapleton House
277-281 Holloway Road
London N7 8HN
UK
Tel: +44 20 7753 3330
Fax: +44 20 7753 3331
Website:
www.tourismconcern.org

Further Information

The newsletters and websites of the membership organizations listed on page 154 are a good source of current information on rainforest conservation. Below is a selection of books which were consulted in the preparation of this book (see also the list on page 154) and which are recommended for further information. Papers published in Oryx, The International Journal of Conservation also provided an important reference source for the preparation of this book.

World Rainforest Movement★
Maldonado 1858-11200
Montevideo
Uruguay
Tel: +598 2403 2989
Fax: +598 2408 0762
Website: www.wrm.org.uy

WRM European Support Office
1c Fosseway Business Centre
Moreton-in-Marsh
Gloucestershire
GL56 9NQ
UK
Tel: +44 1608 652893
Fax: +44 1608 652878

World Wide Fund for Nature International
Avenue de Mont Blanc
CH1196 Gland
Switzerland
Tel: +41 2236 49111
Website: www.wwf.org

WWF-UK★
Panda House
Weyside Park
Godalming
Surrey GU7 1XR
UK
Tel: +44 1483 426444
Fax: +44 1483 426409

WWF-US★
World Wildlife Fund
1250 24th Street N.W.
Washington DC 20037
USA

Intergovernmental Org'ns and Research Institutions

Centre for International Forestry Research (CIFOR)
P.O. BOX 6596 JKPWB
Jakarta 10065
Indonesia
Tel: +62 251 622 622
Website: www.cifor.cgiar.org

CITES Secretariat
International Environment House
15, Chemin des Anémones
CH-1219 Châtelaine Geneva
Switzerland
Tel: +41 22917 8139/40
Website: www.cites.org

Secretariat for Convention on Biological Diversity (CBD)
393, Saint Jacques Street
Suite 300
Montreal Quebec
Canada H2Y 1N9
Tel: +1 514 288 2220
Website: www.biodiv.org

International Tropical Timber Organisation
International Organizations Center
5th Floor
Pacifico-Yokohama 1-1-1
Minato-Mirai Nishi-ku
Yokohama 220-0012
Japan
Tel: +81 45 223 1110
Website: www.itto.or.jp

UNEP-World Conservation Monitoring Centre
219 Huntingdon Road
Cambridge CB3 0DL
UK
Tel: +44 1223 277314
Website: www.unep-wcmc.org

Collins, M.D. Editor (1990) *The Last Rainforests* Mitchell Beazley. Published in association with IUCN: The World Conservation Union.

Daltry, J.C. and Momberg, F. (2000) *Cardomom Mountains* Biodiversity Survey 2000. Fauna & Flora International, Cambridge.

Emanoil, M. Editor (1994) *Encyclopedia of Endangered species* Gale Research Inc., Detroit. Published in association with IUCN: The World Conservation Union.

Harcourt, C.S. and Sayer, J. A. (1996) *The Conservation Atlas of Tropical Forests: The Americas* Simon and Schuster.

Mittermeier, R.A., Myers, N. and Mittermeier, C.G. (1999) *Hotspots: Earth's biologically richest and most endangered terrestrial ecoregions* CEMEX S.A., Conservation International.

Preston-Mafham, K. (1991) *Madagascar: A Natural History* Facts on File Ltd. In association with Survival Anglia.

Smith, N.J.H., Williams, J.T., Plucknett, D.L. and Talbot, J.P. (1992) *Tropical Forests and their Crops* Comstock Publishing Associates (a division of Cornell University Press).

Stattersfield, A.J., Crosby, M.J., Long, A.J. and Wege, D.C. (1998) *Endemic Bird Areas of the World: priorities for biodiversity conservation.* Birdlife International, Cambridge UK.

Whitmore, T. C. (1998) *An Introduction to Tropical Rain Forests* Oxford University Press. Second Edition.

World Commission on Forests and Sustainable Development (1999) *Our Forests, Our Future.* Report of the World Commission on Forests and Sustainable Development. Cambridge University Press.

Davison, G., Cubbitt G. (1998) *The National Parks and Other Wild Places of Malaysia* New Holland Publishers (UK) Ltd. In Association with WWF Malaysia.

Glossary

Araucariaceae: a botanical family of conifer trees, which includes the Monkey Puzzle tree of Chile and Argentina.

Biological diversity: the diversity of ecosystems, species and genetic variation.

Cloud forest: high altitude forest, which is permanently moist because of cloud or mist cover.

Dipterocarpaceae: a botanical family of over 500 South East Asian trees, which are the main source of commercial timber in countries such as Indonesia, Malaysia and the Philippines.

Dipterocarps: trees of the family Dipterocarpaceae.

Epiphyte: a plant that grows on another plant for support but does not extract nourishment from its host. The adjective is epiphytic.

Igapó: a type of Amazonian rainforest that is periodically flooded by blackwater rivers; the black coloration is caused by rotting vegetation.

Kapur: timber produced from trees of Dryobalanops a genus of Dipterocarpaceae.

Lowland rainforest: is the densest and most luxuriant type of rainforest, found in relatively low-lying areas generally at altitudes of up to 900 metres (3,000 feet).

Mangrove forest: a type of tropical rainforest found in sheltered coastal areas. Mangroves are evergreen trees and shrubs that are specially adapted to grow in salty coastal waters.

Meranti: timber produced from trees of Shorea a genus of Dipterocarpaceae.

Myrtaceae: a botanical family of trees and shrubs mainly found in South East Asia, Australia and the Pacific, which includes the genus Eucalyptus.

Swamp forest: inland rainforests found in places with poor drainage or seasonal flooding.

Varzea: a type of Amazonian rainforest that is periodically flooded by whitewater, silt-laden river water.

Verbenaceae: a mainly tropical plant family that includes some mangrove trees and important timber trees such as teak.

Index

A

Aceh Elephant Project 92–3
Africa 28–43
agricultural impact 18
Alaska 148–9
Amazon 118–31
Amazon Manatee 127
Amazon River Dolphin 126
Amazonian Indians 130–1
Ambondrombe forests 51
Andaman Islands 60–1
annual depletion 18
Anolis lizard 115
Arawak Indians 113
Asia, rainforest extent 58
Asian Elephant 90
Asian Paradise-flycatcher 61
Atlantic rainforests 132–9
Atlas Moth 92, 93
avocado 104
Aye-aye 52

B

Baird's Tapir 106
Balanophora sp. 86
bamboo 70, 71
Banana Slug 146, 147
Bataan National Park 94, 95
bats
    Philippine Tube-nosed Fruit-
        Bat 98
    Short-tailed Leaf-nosed Bat
        20, 21
bearded sakis 126
bears
    Honey Bear 122, 123
    Sun Bear 74
Belian 88
Belize 102
Bengal Tiger 74, 75
Big Leaf Maple 149
biodiversity 12
    Amazon 120–1
    Brazil 134
    Cardamom Mountain 67
    Central America 104–5
    Madagascar 46–7, 50–1
    Mount Kinabula National
        Park 78–9
Biosphere Reserves 24
Bird of Paradise
    King 84, 85
    Lesser 84, 85

birds
    conservation, East Africa 40
    threatened species 153
Black Lemur 52, 53
Black Swallowtail Butterfly 84,
    85
Black-faced Lion Tamarin 138
Blue and John Crow Mountains
    114
Blue-crowned Motmot 106,
    107
Boa, Madagascar 50, 51
Bonobo 38–9
Borneo 73
boto 126
Brazil 132–9
Brazilian Rosewood 136–7
Brazilian Three-toed Sloth 138,
    139
Bromeliads 14, 15, 135
Buffalo, Dwarf Water 98
Bukit Tigapuluh National Park
    82, 83
burning 18
    Amazon 124–5
    Indonesia 89
    Tasmania 145
bush viper 33
butterflies
    Black Swallowtail Butterfly
        84, 85
    morpho 135
    Swallowtail Butterfly 98

C

Caboclos 128
Caiman, Spectacled 121
Cambodia 66–7
Cameroon 33
Capuchin monkeys 106
Cardamom Banded Gecko 67
Cardamom Mountains 66–7
Carib Indians 113
Caribbean 110–17
cattle ranches, Amazon 124
CBD *see* Convention on
        Biological Diversity
Cedar, Honduras 105
Celebes Hornbill 79
Central Africa 34–9
Central America 100–9
chameleons 46, 47
    Oustalet's Chameleon 50

chimpanzee 38–9
CI *see* Conservation
        International
CITES *see* Convention on
        International Trade in
        Endangered Species of
        Wild Fauna and Flora
climate regulation 13
cloud forests 12, 23, 47
Coast Redwood 146
Cock of the Rock 124, 125
cocoa 16–17
coffee 16, 17
Conservation International (CI)
    27, 32
conservation measures 22–5
    Amazon 128–9
    Central America 106–7
    Madagascar 50–1
    Philippines 98
    Tasmania 145
conservation organisations
    154–5
Convention on Biological
    Diversity (CBD) 23
Convention on International
    Trade in Endangered
    Species of Wild Fauna and
    Flora, (CITES) 21
Coral-billed Ground-cuckoo 69
Corcovado National Park 108
Costa Rica 108–9
Costa Rican Tarantula 105
Crab-eating Macaque 96
Cracker Butterfly 18
Creeping Fig 59
Crowned Lemur 53
Cuba 116–17
Cuban Solenodon 116–17
Cuban Tree Frog 117
Cuc Phuong National Park
    64–5
cuckoos 69

D

Deer, Lesser Mouse 64, 65
destruction 18–21
    Central America 102–3
    Indonesia 86–9
    Mexico 102–3
dipterocarps
    Indonesia 86
    Philippines 94

rainforest 76, 77
Dog, Small-eared, Amazon 126
dolphins 126
Dominica 112, 113
Drill 37, 38
duiker 40, 41
Dwarf Water Buffalo 98

E

eagles
    Harpy Eagle 127
    Monkey-eating Eagle 98, 99
East Africa 40–1
ebony 86–8
eco-tourism 27
    Africa 42
    Central America 108–9
    Madagascar 50
elephants
    African Elephant 41
    Asian Elephant 64, 90
    hardwood regeneration role
        33
    Sumatra 92–3
Emerald Tree Frog 136, 137
endangered species 21, 138,
    152–3
    Amazon 126–7
    Central Africa 36–9
    Central America 106
    East Africa 40
    Indonesia 90
    Philippines 98
epiphytic ferns 89
evolution
    Madagascar 46–7
    Mentawai islands 82
extractive reserves 128–9

F

farmland 18
Fauna & Flora International
    (FFI) 27, 32, 65, 99, 136
fig trees 59, 71
fires 18
    Amazon 124–5
    Indonesia 89
    Orang-utans 91
    Tasmania 145
fish, Amazon 120
Flame Tree 54
Forest Stewardship Council
    (FSC) 26–7

Fox, Indian Flying 60, 61
frogs
    Cuban Tree Frog 117
    Emerald Tree Frog 136, 137
    Giant Tree Frog 144
    Golden Mantella Frog 46, 47
    Harlequin Frog 108, 109
fruits 16–17
FSC *see* Forest Stewardship
    Council
fuel wood 14
future, Africa 42

G
Gecko, Cardamom Banded 67
genetic resources loss, Indonesia
    · 86
Ghost Mountain
    Cambodia 66, 67
    Madagascar 51
Giant Otter 126, 127
Giant Tree Frog 144
gibbons 63
    Javan Gibbon 90
    White-cheeked Crested
        Gibbon 63
Golden Lion Tamarin 133, 138,
    139
Golden Mantella Frog 46, 47
Golden Parakeet 127
Golden Toad 108–9
gorillas 34, 35
    Mountain Gorilla 37, 38–9
Green Peafowl 84
Green-winged Macaw 137
Groove-billed Toucanet 125
Guan, Horned 106
Gunung Leuser National Park
    80–1, 82, 90, 91

H
Ha Long Bay 62, 63
hamburger connection 103
Harlequin Frog 108, 109
Harpy Eagle 127
hemlock, Alaska 148, 149
Hippopotamus, Pygmy 30, 31
Hoatzin 25
Honduras Cedar, Central
        America 105
Honey Bear 122, 123
Hornbill, Celebes 79
Horned Guan 106
humans 13–17
    Amazon 122–5
    Indian islands 60–1

Madagascar 48–9
Malaysia 73
prehistoric 64
Sarawak 73
South East Asia 59
hunting 35, 124

I
Ibis, Scarlet 115
IFF *see* Intergovernmental
    Forum on Forests
Igapó forests 120, 128, 129
India 60–1
Indian Flying Fox 60, 61
Indonesia 82–93
Indris 52
industrial threats 20
interdependent species 33
Intergovernmental Forum on
        Forests (IFF) 24
Intergovernmental Panel on
        Forests (IPF) 24
international concern 23–5
International Tropical Timber
        Organisation (ITTO) 26
IPF *see* Intergovernmental Panel
        on Forests
Ironwood, Indonesia 88
isolated evolution
    Madagascar 46–7
    Mentawai islands 82
ITTO *see* International Tropical
        Timber Organisation

J
Jaguars, Central America 106,
    107
Jamaica 114–15
Jararaca, pitviper 137
Javan Gibbon 90
Javan Rhinoceros 90

K
Kaka, North Island 143
Karamea Forest 142, 143
Karen people 59
Kayapo Indians 131
King Bird of Paradise 84, 85
Kinkajou 122, 123
Kuang Si Waterfall 68, 69
Kuna Indians 102

L
land-use policy, Malaysia 72
Langur, Red-shanked Douc 63
Laos 68–9

lemurs 46, 52–3
    Black Lemur 52, 53
    Crowned Lemur 53
Lesser Bird of Paradise 84, 85
Lesser Mouse Deer 64, 65
Lesser Slow Loris 64
Liberia 30, 31–2
Lignum vitae, Caribbean 112
limestone ecosystem 62
lion tamarins 138–9
Little Bee-eater 36, 37
lizards
    Anolis 115
    Sailfin Lizard 96, 97
logging 18, 96–7
Long-nosed Whip Snake 60, 61
Long-tailed Macaque 96
Lorikeet, Swainson's 144, 145
Loris, Lesser Slow 64
lowland rainforests 12, 47

M
macaques
    Crab-eating Macaque 96
    Long-tailed Macaque 96
macaws
    Central America 106
    Green-winged Macaw 137
    Scarlet Macaw 137
Madagascar 44–55
Madagascar Boa 50, 51
Magpie, Red-billed Blue 64, 65
mahogany
    Amazon 124
    Caribbean 112
    Central America 105
    Philippines 97
maize 105
Malayan Tapir 74
Malaysia 72–9
Maleo, Sulawesi 92, 93
Man and Biosphere Programme
        24
management
    initiatives 26–7
    Liberian plans 32
    Malaysian timber production
        72
Manatee, Amazon 127
Mangkono, Philippines 97
mangrove forest 12
manioc 130–1
Manu National Park 129, 137
Maple, Big Leaf 149
Marbled Murrelet 146
Mbuti tribe 35

medicial plants 15
Mekong River 68–9
Mentawai islands 82–3
millipedes 124, 125
mining 20, 135
Monkey-eating Eagle 98, 99
monkeys
    Capuchin 106
    Proboscis Monkey 91
    Spider 125
    Squirrel Monkey 24, 25, 139
    White Uakari Monkey 127
    Woolly Spider 139
Monserrat 113
montane rainforests 47, 88, 89
Monteverde Cloud Forest
        Reserve 100–1, 102, 108–9
morpho butterflies 135
Moth, Atlas 92, 93
Motmot, Blue-crowned 106,
    107
Mount Khmaoch 66, 67
Mount Kinabula National Park
    78–9
Mount Sabinyo Volcano 28–9,
    30
Mountain Chicken 113
Mountain Gorillas 37, 38–9
muriqui monkeys 139
Murrelet, Marbled 146
myrtles, Tasmania 144, 145

N
Nicobar Islands 60–1
Nigeria 33
North America 146–9
North Island Kaka 143
Northern Spotted Owl 146,
    147
Nosy Mangabe 47
nuts 16–17

O
oil extraction 20–1
Okapi 36, 37
Okapi Wildlife Reserve 36
Olympic National Park 146,
    147, 148, 149
Operation Amazonia 124
Orang-utans 26, 78, 91
orchids 79, 91
    Madagascar 54–5
    Malaysia 74–5
    Vanilla Orchid 54–5
organizations 154–5
Otter, Giant 126, 127

Oustalet's Chameleon 50
Owl, Northern Spotted 146, 147

P

Palawan 97
palms, Madagascar 54
Pan American Highway 102
Pandrillus Project 38
Paradise-flycatcher, Asian 61
Parakeet, Golden 127
parasitic plants 86
Parrot, St Lucia 117
Pau Brasil 136
Peafowl, Green 84
Philippine Tube-nosed Fruit-bat 98
Philippines, the 94–9
Pink River Dolphin 126
Piranha 121
pitcher plants, Malaysia 74–5
pitviper, Jararaca 137
Poco das Antas Biological Reserve 138
podocarp forest 140–1, 143
pollination, primates 64
Possum, Ringtail 144, 145
prehistoric humans 64
primates
    Brazil 138–9
    Central Africa 36–7, 38–9
    Malaysia 78
    Vietnam 62–5
Proboscis Monkey 20, 91
protected status 26
    Amazon 129
    Malaysia 77–9
pygmies 35
Pygmy Hippopotamus 30, 31

Q

Quetzal, Resplendent 104–5, 106

R

Rafflesia
    R. arnoldii 58, 59
    R. himalayana 67
Rainforest Rescue Programme, Central America 107

rattan industry 14, 74
Red-billed Blue Magpie 64, 65
Red-billed Streamertail 114, 115
Red-eyed Tree Frog 12, 13
Red-shanked Douc Langur 63
Resplendent Quetzal 104–5, 106
rhinoceros
    Javan Rhinoceros 90
    Sumatran Rhinoceros 77, 90, 93
rhododendron 95
Ringtail Possum 144, 145
Rio Conference 23–4
road building 102–3, 124
Rosewood, Bazilian 136–7
rubber 16, 122–3

S

Sabah 72–3, 78
Sailfin Lizard 96, 97
St Lucia Parrot 117
salmon 149
Sarawak 72–3, 77–8
Scarlet Ibis 115
Scarlet Macaw 137
SCNL see Society for the Conservation of Nature in Liberia
seed dispersers 33, 64
Senegal Parrot 33
shifting cultivation 18
    Central Africa 34
    Central America 102–3
    Madagascar 49
    Philippines 94
    Western Ghats 61
Short-tailed Leaf-nosed Bat 20, 21
Siberut 82–3
slash-and-burn agriculture 18
Sloth, Brazilian Three-toed 138, 139
Slug, Banana 146, 147
Small-eared Dog 126
Snake, Long-nosed Whip 60, 61
Society for the Conservation of Nature in Liberia (SCNL) 32

Solenodon, Cuban 116–17
South East Asia 56–79
Spectacled Caiman 121
Spectacled Spiderhunter 71
spider monkeys 125
Spiderhunter, Spectacled 71
spruce, Alaska 148, 149
Squirrel Monkey 24, 25, 139
streamertails 114, 115
subsistence crops 18
sugarcane
    Belize 102
    Brazil 134–5
    Caribbean 112
    Monserrat 113
Sulawesi, Maleo 92, 93
Sumatra 82, 92–3
Sumatran Rhinoceros 77, 90, 93
Sumatran Tiger 18, 19, 82
Sun Bear 74
sustainable management 26–7
    Central America 107
    Siberut 83
Swainson's Lorikeet 144, 145
Swallowtail Butterfly 98
swamp forests 12

T

Taman Negara 72, 73, 76, 77
Tamaraw 98
tamarins 138–9
    Golden Lion Tamarin 133, 138, 139
tapirs
    Baird's Tapir 106
    Malayan Tapir 74
Tarantula, Costa Rican 105
Tasmania 144–5
Teak forests 70
temperate rainforests 140–9
terra firma forests 118–19, 121
Thailand 70–1
threatened species 152–3
tigers
    Bengal Tiger 74, 75
    Sumatran Tiger 18, 19, 82
timber industry 14
    Alaska 148
    Indonesia 86–8
    Malaysia 72

Philippines 96–7
Toad, Golden 108–9
Toco Toucan 134, 135
Tongass National Forest 148
Toucanet, Groove-billed 125
toucans 14, 15
    Toco Toucan 134, 135
Trafalgar Falls, Dominica 112, 113

U

uakaris 126
UK Darwin Initiative 32
United Nations Conference on Environment and Development (UNCED) 23
Urucu fruit 120, 121
Usambara Eagle Owl 40–1

V

Vanilla Orchid 54–5
varzea forests 120, 129
Vietnam 62–5
viper, bush 33
Virgin Jungle Reserves (VJR) 79

W

war protection, Cambodia 66
West Africa 30–3
Western Ghats 60, 61
White Uakari Monkey 127
White-bearded Hermit 20, 21
White-cheeked Crested Gibbon 63
White-faced Tree Duck 49
woolly spider monkeys 139
world distribution map 150–1
World Heritage Convention 24
World Wide Fund for Nature (WWF) 27, 32

Y

Yanomami Indians 130, 131
Yao hill tribe 70, 71
Yekuana Indians 130

Z

Zorro 126

Acknowledgements

Thanks are due to all the people at Fauna & Flora International who provided information or reviewed sections of the text, in particular Mike Appleton, Helene Barnes, Megan Cartin, Jenny Daltry, Georgina Magin, Simon Mickleburgh, Nicky Pulman and Mark Rose.
Mary Cordiner and Val Kapos helped to track down information sources at UNEP-WCMC and their help is also gratefully acknowledged. Special thanks are also due to Peggy Olwell for providing information on the rainforests of North America.

Sara Oldfield

Publisher's Acknowledgements

All photos supplied by The Bruce Coleman Collection except:
Steve Bloom: front cover
Juan Pablo Moreiras / Fauna & Flora International: pages 28, 32, 34, 35, 38, 40, 42, 43, 100, 103, 104, 115 (bottom), 116
Jeremy Holden / Fauna & Flora International: pages 66, 67
Nigel Hicks: pages 94, 96, 99 (top)

Take Action

Everyone can play a part in saving the rainforests. Here are a few ways that you can help.

- Support at least one of the organizations listed on page 154.
- Use wood and paper wisely and recycle both.
- Buy wood products only from forests certified to the standards of the Forest Stewardship Council (FSC).
- Visit a rainforest if you can and travel in an eco-friendly way.